Is War Necessary for Economic Growth?

Is War Necessary for Economic Growth?

Military Procurement and Technology Development

Vernon W. Ruttan

OXFORD
UNIVERSITY PRESS

2006

OXFORD
UNIVERSITY PRESS

Oxford University Press, Inc., publishes works that further
Oxford University's objective of excellence
in research, scholarship, and education.

Oxford New York
Auckland Cape Town Dar es Salaam Hong Kong Karachi
Kuala Lumpur Madrid Melbourne Mexico City Nairobi
New Delhi Shanghai Taipei Toronto

With offices in
Argentina Austria Brazil Chile Czech Republic France Greece
Guatemala Hungary Italy Japan Poland Portugal Singapore
South Korea Switzerland Thailand Turkey Ukraine Vietnam

Published by Oxford University Press, Inc.
198 Madison Avenue, New York, New York 10016

www.oup.com

Library of Congress Cataloging-in-Publication Data
Ruttan, Vernon W.
Is war necessary for economic growth? : military procurement and technology development /
Vernon W. Ruttan.
p. cm.
Includes bibliographical references.
ISBN-13 978-0-19-518804-2

1. Technological innovations—Economic aspects. 2. High technology—Military aspects. 3. Economic development. 4. Defense industries—Economic aspects—United States. 5. Technological innovations—Economic aspects—United States. 6. High technology industries—United States. I. Title.
HC79.T4R878 2006
338.973'0927—dc22 2005007551

The micro processes governing the evolution of military technologies are no different than any other technology.

—Edward W. Constant (2000)

Normal design is very different from radical design such as that confronting the initiators of the turbojet revolution.

—Walter G. Vincenti (2000)

Preface

In this book I examine the impact of military and space-related procurement on the commercial development of six general-purpose technologies. In an earlier book, *Technology, Growth, and Development: An Induced Innovation Perspective* (2001), I discussed several examples but did not give particular attention to the role of military and defense-related procurement as a source of commercial technology development. A major generalization that emerged in my earlier work was that the public sector had played an important role in the research and technology development for almost every industry in which the United States was, in the late twentieth century, globally competitive. During the winter of the 2002–2003 academic year, commitment to present several seminars led to a reexamination of what I had written in *Technology, Growth, and Development*. It became clear to me that defense and defense-related institutions had played a major role in the research and technology development of many of the general-purpose technologies that I had discussed in the earlier book.

The military procurement issue was sitting there in plain sight, but I had been unable or unwilling to recognize it! It was with considerable reluctance, then, that I decided to write this book. I shared the view advanced by John U. Nef in his classic book *War and Human Progress* (1950) that the impact of war on military technology was to due to the intensification of military procurement during wartime, which itself drew on the accumulation of earlier advances in scientific and technical knowledge.

The purpose of this book is to demonstrate that military and defense-related procurement has been a major source of technology development across a broad spectrum of industries that account for an important share of U.S. industrial production. Some colleagues and reviewers have urged me to give more attention to the analytical and policy issues typically included in research in the field of defense research and development. Others have urged me to develop a more comprehensive economic history of military and defense-related technology. My interest is both broader and narrower than these suggestions. My focus on the impact of defense procurement on commercial technology development captures a much more inclusive range of research and technology development than defense research and development. My focus on the impact of defense and defense-related procurement on commercial technology development is narrower, however, than a comprehensive economic history of the development of the military and commercial aircraft or the computer industries.

I owe a very large debt to the numerous colleagues who have critically reviewed earlier versions of the book. (Specific acknowledgments appear at the ends of many chapters.) I owe a particular debt to Richard Nelson, Nathan Rosenberg, and Robert E. Evenson, whose research and counsel have contributed to the development of my own thought on the economics and history of research and technology development. I am indebted to Oxford University Press for permission to draw heavily on several chapters, particularly 7, 9, 10, and 11, from *Technology, Growth, and Development*. Elaine Reber typed and retyped the manuscript, and corrected my spelling and usage. Mary Keirstead's technical editing of the manuscript forced me to clarify both my thoughts and my expression. Louise Letnes has contributed to the accuracy and completeness of citations.

Contents

Is War Necessary for Economic Growth?

1

War and Economic Growth

It is difficult to overemphasize the importance of the historical role that military procurement has played in the process of technology development. Knowledge acquired in making weapons was an important source of the industrial revolution. To bore the condenser cylinders for his steam engines, "Watt had to turn to John Wilkinson, a cannon-borer, who had invented the one machine in all England that could drill through a block of cast iron with accuracy" (Kaempffert 1941, p. 435). In France, the navy provided the market that gave French entrepreneurs an opportunity to catch up with British advances in ferrous metallurgy (McNeill 1982, pp. 177, 211–212). In the United States, what came to be termed *the American system of manufacturing* emerged from the New England armory system of gun manufacture (Rosenberg 1972, pp. 87–116; Smith 1985, pp. 39–86). During almost every year since World War II, defense and defense-related research and technology development expenditures have accounted for at least two thirds of all U.S. federal government research and development (R&D) expenditures (National Science Board 2004, pp. A4, 60).

Historical Perspectives

The relationship between war and economic development has been controversial in economic history.[1] In *Krieg und Kapitalismus* (1913), Werner Sombart argued that in Western Europe war and the preparation for war gave rise to the economic institutions of capitalism (Smith 1985, pp. 29–30). In his classic work, *War and*

1. In this section I have drawn on the exceedingly thorough review of the literature on the history of military institutions and social change by Hacker (1994, pp. 768–834).

Human Progress, John U. Nef argued, partly in response to Sombart's thesis, that the apparent association between war and economic development does not bear up well under careful historical analysis. Impressions that war has an impact on technical change are, for example, formed from the wartime intensification of military procurement that draws on the accumulation of earlier advances in scientific and technical knowledge (Nef 1950, pp. 375–378; Mokyr 1990, pp. 183–186).

This issue became a subject of heated debate in the United States in the 1960s. A study conducted by the Office of the Director of Defense Research and Engineering (HINDSIGHT) purported to show that the significant "research events" that had contributed to the development of twenty major weapons systems were predominantly motivated by military need (Sherwin and Isenson 1967). This view was challenged in studies commissioned by the National Science Foundation and conducted by the Illinois Institute of Technology (*TRACES;* 1968) and Battelle Research Institute (1973). The *TRACES* and Battelle studies adopted a time horizon much longer than the twenty-year period examined in the HINDSIGHT study. These studies concluded that earlier science events, unrelated to military considerations, were of much greater importance relative to technology events as sources of technical change (Mowery and Rosenberg 1979; Thirtle and Ruttan 1987).

During the cold war, defense and defense-related R&D expenditures were criticized as a burden on, rather than as a source of, productivity growth. It was argued that defense and defense-related research and technology development drew scientific and technical capacity away from commercial application and thus slowed technical change in industry (Solo 1962, pp. 49–60; Melman 1974; Kaldor 1981; Dumas 1986; Lichtenberg 1989).[2] In addition, U.S. industry has been criticized for being slow to take advantage of the technology transfer opportunities resulting from military and defense-related R&D. Defense contractors often insulated their military and defense-related R&D from their commercially oriented R&D (Lichtenberg 1989; Alic et al. 1992, pp. 43–44; Markusen and Yudken 1992, pp. 69–100).

In a landmark book published in the mid-1980s, *Military Enterprise and Technological Change*, Merritt Roe Smith, a leading historian of technology, complained that economic historians had largely neglected the contribution of military research,

2. For a thorough review of the literature on military expenditures and economic growth, see Sandler and Hartley (1995). Sandler and Hartley are critical of much of the literature for emphasizing either supply-side or demand-side effects of military procurement expenditures. In attempting to understand the effects of defense expenditures on economic growth, the researcher needs to account for both the supply-side and the demand-side influences (pp. 202–203).

development, and procurement on the development of commercial technology (Smith 1985, pp. 32–37).[3] In my recent book, *Technology, Growth, and Development* (2001), I gave considerable attention to the role of the U.S. public sector in the development of major general-purpose technologies.[4] I discussed several examples but did not give particular attention to the role of military and defense-related procurement as a source of commercial technology development. As aforementioned, although the issue was sitting there in plain sight, I was unable or unwilling to acknowledge it.[5]

The purpose of this book, then, is to demonstrate that military and defense-related R&D and procurement has been a major source of technology development across a broad spectrum of industries that account for an important share of U.S. industrial production. I use the term *military and defense-related procurement* as a shorthand expression to include research, development, and procurement by the National Aeronautics and Space Administration and the Atomic Energy Commission (now the U.S. Department of Energy), and by contractors that conduct R&D in connection with military and defense-related procurement (box 1.1).

I am interested in military and defense-related technology development primarily to the extent that it contributes to commercial technology development. And I am concerned with the impact of military and defense-related technology development and procurement on commercial technology development primarily during the early stages in which commercial technology has most typically drawn on military and defense-related research, technology development, and procurement.[6] As a field of commercial technology that initially drew heavily on military R&D or military and defense-related procurement matures, its dependence on military and defense-related sources tends to decline. The flow of knowledge and technology may then reverse—"from spin-off to spin-on."

Since the end of World War II, the United States has played a predominant role in initiating or implementing the new general-purpose technologies that have

3. A similar criticism could be made of leading scholars in the field of business history and industrial organization (Williamson 1975; Chandler 1977; Porter 1990).

4. General-purpose technologies exert a pervasive impact across a number of industries. Consistent features in the history of general-purpose technologies have been a lengthy period between their emergence and their impact (David 1990; Lipsey, Bekar, and Conlaw 1998) and the cumulating of individual small improvements (Rosenberg 2004).

5. I have not been alone in finessing the role of military procurement in technology development. Members of the scientific community have often preferred to avoid discussion of the intimate relationship between their own research and government investment in military technology (Foreman and Sanchez-Rom 1991).

6. Kira Markiewicz and David Mowery (2004) insist that in many cases the most significant effects of federal spending on industry technology development and diffusion have been the result of procurement rather than of R&D.

Box 1.1. Department of Defense: Unique Science and Technology Development Activities

Basic research	Systematic study directed toward greater knowledge or understanding of the fundamental aspects of phenomena or observable facts, without specific applications toward processes or products in mind.
Applied research	Systematic study to gain knowledge or understanding necessary to determine the means by which a recognized and specific need may be met.
Advanced technology development	Includes all efforts that have moved into the development and integration of hardware for field experiments and tests.
Demonstration and validation	Includes all efforts necessary to evaluate integrated technologies in as realistic an operating environment as possible to assess the performance or cost reduction potential of advanced technology.
Engineering and manufacturing development	Includes those projects in engineering and manufacturing development for service use but which have not received approval for full-rate production.
Research, development, test, and evaluation (RDT&E) management support	Includes research and development (R&D) efforts directed toward support of installations or operations required for general R&D use. Included would be test ranges, military construction, maintenance support of laboratories, operations and maintenance of test aircraft and ships, and studies and analyses in support of R&D programs.
Operational system development	Includes those development projects in support of development acquisition programs or upgrades still in engineering and manufacturing development, but which have received Defense Acquisition Board or other approval for production, or production funds have been included in the Department of Defense (DOD) budget submission for the budget or subsequent fiscal year.

(*continued*)

Box 1.1. *(continued)*

Developmental test and evaluation	Efforts associated with engineering or support activities to determine the acceptability of a system, subsystem, or component.
Operational test and evaluation	Efforts associated with engineering or support activities to determine the acceptability of a system, subsystem, or component.
No separate definition	Major equipment dollars are mixed with the dollars for the "Conduct of R&D" and carried in the RDT&E accounts listed above. In FY 1998, DOD requested a total of $68 million for major R&D equipment.
No separate definition	In FY 1998, close to 90% of the $67 million requested by DOD for R&D facilities was carried separately in military construction accounts. The rest were included in the costs of major development programs and are mixed with the dollars for the "Conduct of R&D" carried in the RDT&E accounts listed above.

Source: Fossum, D. L., S. Painter, V. Williams, A. Yezril, and D. Trinkle. 2000. *Discovery and Innovation: Federal Research and Development Activities in the Fifty States, District of Colombia, and Puerto Rico*. Santa Monica, CA: RAND, p. 615. Reprinted with permission of RAND.

emerged from military and defense R&D and defense-related procurement. Thus, in this book I focus primarily on the United States. I discuss six general-purpose technologies: (1) interchangeable parts and mass production, (2) military and commercial aircraft, (3) nuclear energy and electric power, (4) computers and semiconductors, (5) the Internet, and (6) the space industries.[7] In each chapter I

7. Other areas in which military R&D and military and defense-related procurement have played a significant role in technology development include the laser, radio, food-processing, machine tool, and chemical and medical industries. I discuss the role of military demand on technical change in the chemical industry in some detail in Ruttan (2001, pp. 286–315). In this book I do not discuss the large number of secondary spin-offs from military and defense-related research and procurement. A classic example is the microwave oven, a spin-off from research on radar, developed by Raytheon. Popular accounts of spin-offs from military and space research and pro-

speculate, in the spirit of counterfactual analysis, on whether, in the absence of the stimulus resulting from military procurement, commercial development would have occurred "anyway"—or at least more slowly.[8] In a concluding chapter I address the role of military and defense-related procurement on technology development in the United States in a postindustrial economy.

Rate and Direction of Technical Change

In this section I present an overview of the theoretical perspectives that have guided economists in their attempts to understand the economic forces that have influenced the rate and direction of technical change in commercial technology development. Throughout the book, I will explore the extent to which these theoretical perspectives illuminate the sources of military and defense-related technologies.

Historians of science and technology, and scientists and engineers themselves, have traditionally sought to interpret advances in scientific and engineering knowledge internally—that is, in terms of the motives of individual scientists and engineers or in terms of the culture of scientific and engineering societies and communities, rather than in terms of changes or differences in social, political, and economic environments.[9] Internalist interpretations have become considerably less compelling as advances in scientific and engineering knowledge have increasingly emerged from large government and industrial laboratories and from contract research carried out at major research universities. These interpretations have nevertheless retained substantial currency in military history (Hacker 1994). Edward Constant (2000a) has insisted, however, that "the micro processes governing the

curement have often been excessively extravagant or wholly fictitious (Alic et al. 1992, pp. 54–81).

8. Counterfactual analysis became a central concern in the "new economic history" introduced by a group of younger economic historians in the late 1950s and 1960s. As practiced by the new economic historians, counterfactual analysis involved the application of economic theory and econometric method to establish the effects of technical and institutional innovations on the course of economic development. The net effects of technical and institutional change on development involve "a comparison between what actually happened and what would have happened in the absence of the specified circumstances" (Fogel 1966, p. 653). The introduction of formal analytical-quantitative methodology became a source of substantial debate among economic historians in the 1960s (Fogel 1967). The counterfactual arguments in this book are based on traditional narrative rather than on econometric analysis.

9. For an extreme example, see David Noble, *The Religion of Technology* (1997). Noble argues that since the Middle Ages Christianity has been a dominant inspiration for advances in technical knowledge. But in the modern world "the other worldly preoccupations of later day spiritual men . . . have contributed enormously to the world arsenal for warfare, surveillance and control" (p. 206).

evolution of military technologies are no different than any other technology" (p. 288).[10]

During the 1960s through the 1980s, economists developed many new theories and insights into the sources of technical change. In the 1960s and 1970s, they focused their attention on the development and testing of the theory of induced technical change, particularly on the role of economic forces—primarily changes in demand and changes in relative factor prices—on the rate and direction of technical change. In the late 1970s and early 1980s, attention shifted to evolutionary models inspired by a revival of interest in Joseph Schumpeter's (1934) work on the sources of economic development. Beginning in the early 1980s, these theories were supplemented by the development of historically grounded "path-dependent" models of technical change. Each has contributed substantial insight into the generation and choice of new technology.

Induced Technical Change

In a now classic study of patent statistics, Jacob Schmookler (1966) showed that in the United States during much of the nineteenth century, when investment rose, capital goods inventions also rose; and when investment declined, the flow of patents declined. His intensive historical studies of a series of major inventions led him to conclude that demand was a more important source of change in the *rate* of technical change than advances in the state of knowledge (Schmookler 1962; Schmookler 1966).[11]

Arguments about the relative importance of demand-side and supply-side forces intensified in the late 1960s. I referred earlier to a study conducted by the Office of the Director of Defense Research and Engineering that purported to show that significant "research events" that contributed to the development of twenty major weapons systems were motivated primarily by military "need" rather than by disinterested scientific inquiry (Thirtle and Ruttan 1987, pp. 6–11). Subsequent studies have shown that technical changes induced by both supply- and demand-side factors have played an important role in the life cycles of many industries (Walsh 1984, p. 233). But there should be no argument that growth in demand does represent a powerful inducement to the allocation of resources to research, and that

10. I have reviewed the evolution of thought on the sources of innovation in scientific and technical knowledge in Ruttan (2001, pp. 63–99, 534–538) and Ruttan (2003, pp. 82–87), and on the rate and direction of technical change in Ruttan (2001, pp. 100–146).

11. The Schmookler work initiated an intense debate among economic historians on the relative importance of "demand pull" relative to "supply push" as a source of technical change (Mowery and Rosenberg 1979).

military procurement has been an important source of demand-induced technical change.

Interest in the effects of changes (and differences) in relative factor endowments and prices on the *direction* of technical change was initially stimulated by an observation by Sir John Hicks: "The real reason for the predominance of labor saving innovation is surely that . . . a change in relative prices of factors of production is itself a spur to innovation and to innovation of a particular kind—directed at economizing the use of a factor which has become relatively expensive" (1963, pp. 124–125).[12]

The first econometric tests of the microeconomic theory of induced technical change were conducted by Yujiro Hayami and Vernon W. Ruttan against the history of technical change in agriculture in the United States and Japan (Hayami and Ruttan 1973; Hayami and Ruttan 1985). It was apparent that the enormous changes in land-labor ratios over time in the two countries could not be explained by simple factor substitution. Hayami and I showed that (1) land and mechanical power were complements, and mechanical power and labor were substitutes; and (2) fertilizer and land infrastructure were complements, and fertilizer and land were substitutes. Our econometric tests confirmed that the enormous changes in factor proportions that occurred in the process of development in the two countries represented a process of dynamic factor substitution in response to technical change induced by changes in relative factor endowments. Japan initially followed a "biological," and the United States a "mechanical," technical trajectory. Since the middle of the twentieth century, the trajectories have experienced substantial convergence.

The Hayami-Ruttan work was followed by a large number of empirical tests of the microeconomic version of the induced-technical-change hypothesis in the agricultural, natural resource, and industrial sectors (Thirtle and Ruttan 1987). The results of these tests confirmed that changes (and sometimes differences) in relative factor endowments and prices exerted a pervasive impact on the direction of technical change. The only formal test of the induced-innovation hypothesis against military procurement is Ames and Rosenberg's (1968) study of technical change in the Springfield Armory (United States) and the Enfield Arsenal (United King-

12. The theory of induced innovation, particularly its macroeconomic variant, was the subject of considerable controversy in the mid-1960s. For a review, see Nordhaus (1973). See also the criticism by Mokyr (2002, pp. 292–294). The macroeconomic variant of induced technical change has since the early 1990s been discussed under the rubric of endogenous technical change (Romer 1986; Lucas 1988; Ruttan 1998). Rosenberg (2004) has argued that the growth of scientific and technical knowledge has become far more endogenous over the course of the twentieth century.

dom). They found that part of the explanation for the higher labor productivity at Springfield was to be found in the relative prices of raw materials (wood and metal) and in the wages of highly skilled workers compared to less skilled workers. It is hard to believe that the enormous rise in the price of labor compared to the price of capital equipment in the U.S. economy has not played a significant role in inducing the capital intensity of U.S. military technology. Only an economy that places an extremely high value on human capital could devote resources to producing a tool as expensive as stealth aircraft.[13]

Evolutionary Theory

Modern interest in an evolutionary theory of technical change derives largely from the work of Richard R. Nelson and Sidney Winter in the 1970s. They followed a series of articles with the highly acclaimed book *An Evolutionary Theory of Economic Change* (Nelson and Winter 1982).[14] Evolutionary theory jettisons much of what Nelson and Winter consider the excess baggage of the neoclassical theory, including profit-maximizing firm behavior and the production function as a description of firm-level technology, which play a central role in the theory of induced technical change. The theoretical cornerstone of the Nelson-Winter model is the behavioral theory of the firm in which profit-maximizing behavior is replaced by the concept of "routine"—a term that includes characteristics ranging from well-specified technical *procedures* for producing things and for managing practices, research, and development, to business strategies about product diversification, investment, and marketing (Nelson and Winter 1982, p. 14).

The two fundamental mechanisms in the several Nelson-Winter models are the *search* for better techniques and the *selection* of firms or technologies by the market. The learning activities leading to technical change are characterized by (1) local search for technical innovations, (2) imitation of practices employed by other firms, and (3) "satisfying" economic behavior. Search for new technology, whether generated internally by R&D, or transferred from suppliers or competitors, is set in motion when profits fall below a certain threshold. A change in factor prices, or

13. In the United States, military force changes since the early 1970s have been associated with very substantial substitution of capital for labor. Between 1970 and 2003 the equipment cost per person approximately doubled and the budget cost per person increased by about two thirds in the U.S. armed forces: "On average the United States spends just short of $300,000 per person in the armed forces—twice as much as its closest allies and far more than any potential antagonists" (Deitchman 2004, p. 63).

14. Evolutionary theory in economics has experienced substantial criticism and elaboration since its early articulation by Nelson and Winter (1982). See, for example, the essays in Ziman (2000).

a rise in the wage rate, for example, will cause some techniques to fail the profitability test, and different techniques to pass the tests they might have failed at a lower wage rate.

History plays an important role in the Nelson-Winter evolutionary models. The condition of the industry in each time period shapes its condition in the following period: "Some economic processes are conceived as working very fast, driving some of the model variables to temporary equilibrium values within a single period. . . . Slower working processes of investment and technological and organizational change operate to modify the data of the short run equilibrium system from period to period or even from instant to instant" (Winter 1984, p. 290).

The Nelson-Winter evolutionary theory has provided substantial insight into the operation of large bureaucratic firms, such as those in the aerospace industry, particularly in their approach to technical innovation. The pipeline model of military procurement (see chapter 8) is not unlike the process assumed in some of the Nelson-Winter simulations. I have not, however, been able to identify any empirical studies of the rate and direction of military or defense-related R&D that have attempted to test the Nelson-Winter evolutionary theory.

Path Dependence

The argument that technological change is *path dependent* was vigorously advanced by W. Bryan Arthur and several colleagues beginning in the early 1980s (Arthur 1983; Arthur 1989; Arthur 1994). In the middle and late 1980s, Paul A. David presented the results of a series of historical studies—of the typewriter keyboard, the electric light and power industry, and others—that served to buttress the plausibility of the path-dependence perspective (David 1985; David and Bunn 1988).

The distinctive contribution of the work by Arthur, David, and their colleagues was their emphasis on increasing returns to scale as a source of technical "lock-in." In some nonlinear dynamic systems, positive feedbacks, termed *polya* processes, may cause particular technological patterns or structures to become self-reinforcing: "Often there is a multiplicity of patterns that are candidates for long term self-reinforcement. A combination of small events early in the R&D process pushes the dynamics of technical choice into the orbit of one of these paths and thus 'selects' the structure that the system eventually locks into" (Arthur 1994, p. 294). Arthur cites the almost accidental dominance of personal computer software by Microsoft as an example (1989, p. 127).

David has presented the dominance of the QWERTY (the first six letters on the left top row) typewriter and computer keyboard as a particularly compelling example of how an inefficient (from today's perspective) technology was introduced and has persisted. David's explanation of why this occurred is that an innovation in typing method, touch typing, gave rise to three features that caused QWERTY to become "locked in" as the dominant keyboard arrangement. These features were *technical interrelatedness*, *economies of scale*, and *quasi-reversibility* of investment (David 1985, p. 334). *Technical interrelatedness* refers to the need for system compatibility—in this case the linkage between the design of the typewriter keyboard and the typists' memory of a particular keyboard arrangement. *Economies of scale* refers to the decline in the user cost of the QWERTY system (or any other system) as it gains acceptance relative to other systems. *Quasi-reversibility* is the result of the acquisition of specific touch typing skills (the "software"). These characteristics are sometimes bundled under the rubric of "network externalities."

The development of nuclear power for commercial use (see chapter 4) is an example of path dependence drawn from defense-related procurement. In the early 1950s the Atomic Energy Commission (AEC) initiated a program to support and evaluate alternative nuclear reactor designs. Before this evaluation was complete, events conspired to force a choice of the light water reactor. One of these events was the choice by the U.S. Navy of the light water reactor for propulsion of its nuclear submarine. A second was President Dwight D. Eisenhower's desire for early implementation of nuclear power generation as a showpiece for his Atoms for Peace initiative. A third was the subsidies to General Electric and Westinghouse that enabled them to enter the international market with turnkey light water reactors in time to preempt the alternatives being pursued by other U.S. firms and national governments.

Radical Technology

The six general-purpose technologies that I discuss in this book can all be described as radical or revolutionary technologies. They all represented revolutionary departures from existing technological trajectories. While the three economic models just discussed provide substantial insight into the rate and direction of incremental changes in technology, they do not address the sources of revolutionary new technologies.

An earlier generation of historians of technology viewed major inventions as a result of transcendental insight—as due to the unique inspiration of the occasional genius who achieves advances in knowledge through the exercise of intuition (Ruttan 2001, pp. 65–66). In a landmark book on the history of the turbojet revolution, Edward Constant (1980) advanced the concept of *presumptive anomaly* as a source of radical advances in technology: "Presumptive anomaly occurs in technology, not when the conventional system fails in any absolute or objective sense, but when assumptions derived from science indicate either under some future conditions the conventional system will fail (or function badly) or that a radically different system will do a better job" (Constant 1980, p. 15; see also Constant 2000b).[15] Thus, in the case of the turbojet, insight derived from aeronautics in the 1920s created a presumption among a few aircraft engineers that, over the longer term, fundamental constraints would be confronted on the performance of the piston-propeller system of aircraft propulsion (see chapter 3). Another example is the realization in the late 1930s by Dr. Mervin Kelly, director of research at Bell Telephone Laboratories, that the heat-generation capacity of vacuum tubes would become a constraint on the development a more rapid telephone switching technology (see chapter 5).

It is not necessary that the insight that gives rise to a perception of anomaly be derived from science. Advances in engineering or agronomic knowledge may also give rise to presumptive anomaly. When a radical new engineering technology is envisaged, it may be initially judged as less efficient than the system it is designed to replace. Furthermore, a radical new general-purpose technology will generally, over time, do much more than perform existing functions more efficiently. It will make possible new functions that the technology it replaces could not perform (Aitken 1985, pp. 7–12) Thus, the electronic digital computer not only replaced tabulating and calculating machines, but also opened up the possibility of entirely new communications technologies (see chapter 5).

The presumptive anomaly and the three economic models of the sources of technical change are elements that may in the future be incorporated into a more comprehensive or general model of the sources of technical change. But such a model does not yet exist. There is yet no adequate general interpretation of the sources of defense and defense-related technical change (Rosen 1991, pp. 1–53). Each has been useful in advancing our understanding of the sources of technical change. In this book I draw on the several models as I attempt to understand the role of military R&D and defense and defense-related procurement on commercial technology development.

The Book Plan

In the following chapters of this book I discuss the role of military and defense-related R&D procurement on the development of six general-purpose technologies

15. The Constant presumptive anomaly model bears some resemblance to the theory of paradigm shifts advanced by Kuhn to explain discontinuities in the history of scientific theory (1970).

that played a decisive role in the development of the U.S. economy in the twentieth century.

In chapter 2, I discuss the role of military procurement in the development in the early decades of the nineteenth century of what came to be known as *the American system of manufacturing.* Economic historians have characterized the American system, or more appropriately the New England armory system, as the assembly of complex products produced from individual interchangeable parts. Its first important application was in the manufacture of firearms at the U.S. Army Harpers Ferry (Virginia) and Springfield (Massachusetts) armories. During the second half of the nineteenth century, "armory practice" diffused to other branches of manufacturing, such as sewing machine and bicycle manufacture. It emerged in its most highly developed form in the mass production of automobiles at the Ford Motor Company in the first two decades of the twentieth century.

In chapter 3, I describe the intimate relationship between military and commercial aircraft development. The aircraft industry was unique among manufacturing industries in that a government research organization, the National Advisory Committee for Aeronautics (NACA), was established shortly before U.S. entry into World War I to conduct research on military and commercial aircraft technology and design. NACA was an important, and efficient, source of new knowledge and new technology for the military and commercial aircraft industries for four decades, until it was absorbed into the National Aeronautics and Space Administration (NASA) in 1958. NASA continues to be involved in almost every aspect of aircraft research and technology development.

In chapter 4, I trace the origin of the nuclear power industry to the World War II Manhattan Project, which was organized to develop and build the atomic bomb. The demonstration of controlled nuclear fission at the University of Chicago's Stagg Field on December 2, 1942, initiated the chain of events that led to the development of nuclear power. Since the beginning it has not been possible to understand the development of the nuclear power industry apart from the application of nuclear energy in military technology. The design of the first U.S. nuclear power reactor, located at Shippingport, Pennsylvania, was adapted from nuclear reactors developed in the early 1950s to power nuclear submarines. This early commitment to the light water reactor design appears in retrospect to have been at least a partial source of the failure of the nuclear power industry to realize the promise that it appeared to have in the 1950s.

In chapter 5, I describe the development of the computer industry from the first all-purpose electronic digital computer developed by John W. Mauchly and J. Prosper Eckert and their associates at the University of Pennsylvania's Moore School of Electrical Engineering with funding from the U.S. Army Ballistic Research Laboratory. The first working transistor emerged from the solid-state research led by William Shockley, John Bardeen, and Walter Brattain at Bell Laboratories in the late 1940s. The transition between the initial development of the transistor and the subsequent development of military and commercial application in the 1950s was substantially funded by the Army Signal Corps.

In chapter 6, I trace the role of the Defense Advanced Research Projects Agency (ARPA), from the initial interest of its Information Processing Techniques Office in man-machine interaction in the early 1960s; to the development of a project to interconnect large computers at a number of academic, industrial, and government computer centers in the early 1970s; through the invention of the Internet and its eventual privatization. It was not until 1990 that military responsibility for the Internet was finally terminated.

In chapter 7, I explore the implications of the Space Act of 1958 that established NASA to ensure U.S. leadership in space technology—including weather, communications, and remote-sensing satellites—considered important to national prestige, national defense, and foreign policy. NASA played an early entrepreneurial role in the development of both communications and earth-observing satellites for military and commercial purposes. In the past several decades, development of civil applications were slowed by national security considerations and by ideologically burdened privatization policies that threatened economic viability.

In the final chapter, I address several issues that bear on the impact of military and defense-related procurement on technology development in the United States in the future. One is whether changes in the structure of the American economy preclude military procurement from playing an important role in the development of advanced technology in the future, one comparable to that it played in the past. It has been argued that in many areas of technology the direction of spin-off in recent decades has shifted to one from commercial to military technology, rather than one from military to commercial technology. A second, related argument is that the military and defense-related industries have become so small, compared to the size of the U.S. economy, or even to the manufacturing sector, that it no longer exerts significant leverage on the rate or direction of technical change (Alic et al. 1992; Nelson and Wright 1992).

If military or defense-related procurement does remain an essential source of advanced technology development, then a second, more disturbing question comes to mind. Will war, or at least the threat of war, continue to be necessary to induce the "political will" to mobilize the scientific, technical, and financial resources to generate new general-purpose technologies? I will return to this issue in the final chapter.

REFERENCES

Aitken, H.G.J. 1985. *The Continuous Wave: Technology and American Radio, 1900–1932*. Princeton, NJ: Princeton University Press.

Alic, J. A., M. Branscomb, H. Brooks, A. B. Carter, and G. L. Epstein. 1992. *Beyond Spinoff: Military and Civilian Technology in a Changing World*. Boston: Harvard Business School Press.

Ames, E., and N. Rosenberg. 1968. "The Enfield Arsenal in Theory and History." *Economic Journal* 78: 827–842.

Arthur, W. B. 1983. "On Competing Technologies and Historical Small Events: The Dynamic of Choice under Increasing Returns." International Institute of Applied Systems Analysis Paper WP 83–90, Laxenburg, Austria.

———. 1989. "Competing Technologies, Increasing Returns, and Historical Small Events." *The Economic Journal* 99: 116–131.

———. 1994. *Increasing Returns and Path Dependence in the Economy*. Ann Arbor: University of Michigan Press.

Battelle Research Institute. 1973. *Interaction of Science and Technology in the Innovation Process: Some Case Studies*. Columbus, OH: Battelle Research Institute.

Chandler, A. D., Jr. 1977. *The Visible Hand: The Managerial Revolution in American Business*. Cambridge, MA: Harvard University Press.

Constant, E. W. 1980. *The Origins of the Turbojet Revolution*. Baltimore: Johns Hopkins University Press.

———. 2000a. "The Evolution of War and Technology." In *Technological Knowledge as an Evolutionary Process*, ed. J. Ziman. Cambridge: Cambridge University Press, 281–298.

———. 2000b. "Recursive Practice and the Evolution of Technological Knowledge." In *Technological Knowledge as an Evolutionary Process*, ed. J. Ziman. Cambridge: Cambridge University Press, 219–233.

David, P. A. 1985. "Clio and the Economics of QWERTY." *American Economic Review* 76: 332–337.

———. 1990. "Computer and Dynamo: A Historical Perspective on the Modern Productivity Paradox." *American Economic Review* 80: 355–361.

David, P. A., and J. A. Bunn 1988. "The Economics of Gateway Technologies and Network Evolution: Lessons from Electric Supply History." *Information Economics and Policy* 3: 165–202.

Deitchman, S. J. 2004. "Completing the Transformation of U.S. Military Forces." *Issues in Science and Technology* 20: 62–68.

Dumas, L. J. 1986. *The Overburdened Economy*. Berkeley: University of California Press.

Fogel, R. W. 1966. "The New Economic History: Its Findings and Methods." *Economic History Review* 19: 642–656.

———. 1967. "The Specification Problem in Economic History." *Journal of Economic History* 27: 283–308.

Foreman, P., and J. M. Sanchez-Rom. 1991. Introduction. *National Military Establish-*

ments and the Advancement of Science and Technology: Studies in Twentieth-Century History, ed. P. Foreman and J. M. Sanchez-Rom. Dordrecht, Netherlands: Kluwer Academic, ix–xv.

Hacker, B. C. 1994. "Military Institutions, Weapons and Social Change: Toward a New History of Military Technology." *Technology and Culture* 35: 768–834.

Hayami, Y., and V. W. Ruttan. 1973. 1985. *Agricultural Development: An International Perspective*. Baltimore: Johns Hopkins University Press.

Hicks, J. 1963 [1932]. *The Theory of Wages*, 2nd ed. London: Macmillan.

Illinois Institute of Technology. 1968. *Technology in Retrospect and Critical Events in Science (TRACES)*. Chicago: Illinois Institute of Technology. Mimeograph.

Kaempffert, W. 1941. "War and Technology." *American Journal of Sociology* 46: 431–444.

Kaldor, M. 1981. *The Baroque Arsenal*. New York: Hill & Wang.

Kuhn, T. S. 1970. *The Structure of Scientific Revolutions*. Chicago: University of Chicago Press.

Lichtenberg, F. R. 1989. "The Impact of the Strategic Defense Initiative on U.S. Civilian R&D Investment and Industrial Competitiveness." *Social Studies of Science* 19: 265–282.

Lipsey, R. G., C. Bekar, and K. Conlaw. 1998. "What Requires Explanation?" In *General Purpose Technologies and Economic Growth*, ed. E. Helpman. Cambridge, MA: MIT Press, 16–54.

Lucas, R. E., Jr. 1988. "On the Mechanics of Economic Development." *Journal of Monetary Economics* 22: 3–42.

Markiewicz, K., and D. C. Mowery. 2004. "The Federal Role in Financing Major Inventions: Information Technology during the Postwar Period." Berkeley: Haas School of Business, University of California. Mimeograph.

Markusen, A., and J. Yudken. 1992. *Dismantling the Cold War Economy*. New York: Basic Books/Harper Collins.

McNeill, W. H. 1982. *The Pursuit of Power: Technology, Armed Forces and Society since A.D. 1000*. Chicago: University of Chicago Press.

Melman, S. 1974. *The Permanent War Economy: American Capitalism in Decline*. New York: Simon & Schuster.

Mokyr, J. 1990. *The Lever of Riches: Technology, Creativity, and Economic Progress*. New York: Oxford University Press.

———. 2002. *The Gifts of Athena: Historical Origins of the Knowledge Economy*. Princeton, NJ: Princeton University Press.

Mowery, D., and N. Rosenberg. 1979. "The Influence of Market Demand upon Innovation: A Critical Review of Some Recent Empirical Studies." *Research Policy* 8: 102–153.

National Research Council. 1999. *Funding a Revolution: Government Support for Computing Research*. Washington, DC: National Academy Press.

———. 2005. *Measuring Research and Development Expenditures in the U.S. Economy*. Washington, DC: National Academies Press.

National Science Board. 2004. *Science and Engineering Indicators. 2004*. Arlington, VA: National Science Foundation, NSB 04–1A.

Nef, J. U. 1950. *War and Human Progress: An Essay on the Rise of Industrial Civilization.* Cambridge, MA: Harvard University Press.

Nelson, R. R., and S. G. Winter. 1982. *An Evolutionary Theory of Economic Change.* Cambridge, MA: Harvard University Press.

Nelson, R. R., and G. Wright. 1992. "The Rise and Fall of American Technology and Leadership: The Postwar Era in Historical Perspective." *Journal of Economic Literature* 23: 1931–1964.

Noble, D. F. 1997. *The Religion of Technology: The Divinity of Man and the Spirit of Invention.* New York: Knopf.

Nordhaus, W. D. 1973. "Some Skeptical Thoughts on the Theory of Induced Innovation." *Quarterly Journal of Economics* 87: 208–219.

Porter, M. E. 1990. *The Competitive Advantage of Nations.* New York: Free Press.

Romer, P. M. 1986. "Increasing Returns and Long-Run Growth." *Journal of Political Economy* 94: 1002–1037.

Rosen, S. P. 1991. *Winning the Next War: Innovation in the Modern Military.* Ithaca, NY: Cornell University Press.

Rosenberg, N. 1972. *Technology and American Economic Growth.* New York: Harper & Row.

———. 2004. "Endogenous Forces in Twentieth-Century America." Stanford, CA: University of Stanford, Department of Economics. Mimeograph.

Ruttan, V. W. 1998. "The New Growth Economics and Development Economics." *Journal of Development Studies* 35: 1–26.

———. 2001. *Technology, Growth, and Development: An Induced Innovation Perspective.* New York: Oxford University Press.

———. 2003. *Social Science Knowledge and Economic Development: An Institutional Design Perspective.* Ann Arbor: University of Michigan Press.

Sandler, T., and K. Hartley. 1995. *The Economics of Defense.* Cambridge: Cambridge University Press.

Schmookler, J. 1962. "Changes in Industry and in the State of Knowledge as Determinants of Industrial Invention." In *The Rate and Direction of Inventive Activity: Economic and Social Factors,* ed. R. R. Nelson. Princeton, NJ: Princeton University Press, 195–232.

———. 1966. *Innovation and Economic Growth.* Cambridge, MA: Harvard University Press.

Schumpeter, J. 1934 [1912]. *The Theory of Economic Development,* trans. R. Opie from the revised 1926 German edition. Cambridge, MA: Harvard University Press.

Sherwin, C. W., and R. S. Isenson. 1967. "Project HINDSIGHT: A Defense Department Study of the Utility of Research." *Science* 156: 1571–1577.

Smith, M. R. 1985. "Army Ordinance and the 'American System' of Manufacturing, 1851–1861." In *Military Enterprise and Technological Change: Perspectives on the American Experience,* ed. M. R. Smith. Cambridge, MA: MIT Press, 39–86.

Solo, R. A. 1962. "Gearing Military R&D to Economic Growth." *Harvard Business Review* 40: 49–60.

Sombart, W. 1913. *Krieg und Kapitalismus.* Munich: Dunker and Humblot.

Thirtle, C. G., and V. W. Ruttan. 1987. *The Role of Demand and Supply in the Generation and Diffusion of Technical Change.* London: Harwood Academic.

Walsh, V. 1984. "Invention and Innovation in the Chemical Industry: Demand-Pull or Discovery-Push." *Research Policy* 13: 211–234.

Williamson, O. E. 1975. *Markets and Hierarchies.* New York: Free Press.

Winter, S. 1984. "Schumpeterian Competition in Alternative Technological Regimes." *Journal of Economic Behavior and Organization* 5: 287–320.

Ziman, J., ed. 2000. *Technological Innovation as an Evolutionary Process.* New York: Cambridge University Press.

2

Interchangeable Parts and Mass Production

In 1800 the manufacturing sector accounted for less than 10 percent of U.S. commodity production. By the end of the century, it accounted for over 50 percent. A unifying theme of this dramatic growth in the manufacturing sector was what came to be referred to as "the American System of Manufacturing" (Hounshell 1984, pp. 331–336). Economic historians have traditionally characterized the American system as the assembly of complex products from interchangeable individual parts (Rosenberg 1972, pp. 87–116). The system was developed for the manufacture of guns for the Army at the national armories at Springfield, Massachusetts, and Harpers Ferry, Virginia (Smith 1977).[1]

My primary purpose in this chapter is to discuss the American, or more appropriately, the New England, armory system to illustrate that the role of military procurement in the development of general-purpose technology in the United States extends back to the early years of the nation. The New England armory system introduced during the first half of the nineteenth century had a pervasive impact on the development of American manufacturing. It was the precursor of what has been termed the "Fordist" system of mass production. A second theme of this chapter is the difficulty of achieving rapid technology transfer even when the participants are working under a unified command and control system or are in the same industry.

1. The conceptual basis for a system of uniform ordinance manufacture based on interchangeable parts was initially developed in France. It was diffused to the United States by French officers during the Revolutionary War. The continuing puzzle of why a machine process for production of guns was first developed in the United States rather than in France remains unresolved (Smith 1985, pp. 39–86). Ames and Rosenberg (1968) suggest an induced technical change interpretation of the prior adoption of machine methods for producing muskets in the United States than in Britain.

Interchangeable Parts

The significance of interchangeability can best be understood when compared to the handicraft technology previously used in gun making.[2] Handicraft gun making involved precisely fitting together, primarily by hand filing, individual components produced by a large number of craftsmen. Substantial skill and patience were required for tasks such as filing and recessing the gunstock to properly accommodate the lock and barrel and correctly arranging the pin and screws. In contrast, the system of interchangeability required less skill and thus vastly simplified gun production, repair, and maintenance. It also meant that an army in the field no longer had to be accompanied by armorers to repair a broken part or fit a new part (Mokyr 1990, pp. 136–137).

Springfield and Harpers Ferry

Before 1797 the U.S. War Department purchased its arms, whether imported or produced domestically, from private contractors. In 1794 President George Washington, disturbed by the inadequate performance and corruption of the contract system, proposed a bill, which the Congress passed, to create up to four public arsenals and magazines to manufacture and supply arms to the U.S. Army Department. The bill authorized the president to decide on the locations and to select (and dismiss) the armory superintendents. A site at Springfield, Massachusetts, already owned by the government, was selected shortly after the bill was passed. It was not until 1798, after considerable controversy, that a second site, at Harpers Ferry, was selected (Smith 1977, pp. 28–32).

In 1812 a U.S. Ordnance Department was established as an agency for the inspection and distribution of military supplies, and it was given jurisdiction for the Springfield and Harpers Ferry armories. Its first director, Colonial Decius Wadsworth, staffed the department with able young military engineers. His chief assistant and successor, Colonel George Bomford, was given responsibility for overseeing and promoting greater efficiency in arms manufacture at the Springfield and Harpers Ferry armories (Smith 1977, p. 106). In the decade after the War of 1812, additional armories were established. Several of the arsenals, such as the Springfield Arsenal and the Frankford Arsenal (located outside of Philadelphia), played an important role in the industrial development of the regions in which they were located.

Traditionally the manufacture of rifles and pistols involved a number of separate

2. In this section I draw substantially on Smith (1977) and Ruttan (2001, pp. 426–428).

specialized branches of labor: barrel making, lock forging, lock filing, brazing, stocking, finishing, and assembly: "Despite the rudimentary division of labor involved in the manufacturing process each gun remained a handicraft product" (Smith 1977, p. 79). When Roswell Lee was appointed superintendent at Springfield in 1815, he initiated a series of technical and managerial innovations designed to make the Springfield Armory one of the most advanced manufacturing establishments in the United States. At the time he assumed the position of superintendent, the occupational specialties had risen from 5 to 34. They rose to 68 in 1820, and to over 100 in 1825. The division of labor at the Harpers Ferry Armory lagged behind that at Springfield, reaching 64 occupational specializations in 1825. Except for brief intervals, a combination of inept, and sometimes corrupt, management and resistance to division of labor by the skilled armorers delayed the transition to more complete mechanization of operations at Harpers Ferry (Smith 1977, pp. 79–83).[3]

One of the most remarkable pieces of equipment introduced at the Springfield Armory was a lathe for producing gunstocks, invented by Thomas Blanchard. The invention involved the difficult task of designing a machine that could turn out irregular forms. Superintendent Lee arranged for Blanchard to become an "inside contractor." The terms of the contract called for the armory to furnish "shop space, water privileges, raw materials, and general use of tools and machinery of the armory." Moreover, "Blanchard agreed to provide his patented machinery royalty free and to hire his own workmen" (Smith 1977, p. 135). He was paid thirty-five cents each for the musket stocks he produced. The lathe and other machines that Blanchard invented eliminated the need for skilled labor in one of the major divisions of gun production.

As early as 1811 John H. Hall, then a proprietor of a woodworking establishment in Portland, Maine, had developed a prototype of a breech-loading rifle. His attempts to patent the gun were not resolved until he agreed to share patent rights with the commissioner of patents, who himself attempted to claim prior invention. Through the intervention of influential political friends from Maine and the Navy, Hall was able to bring his invention, now substantially improved, to the attention of the War Department. Between 1813 and 1819 he received several small orders for his breech-loaded rifles. A test by the Artillery School concluded that Hall's

3. "While the arms manufactured at Harpers Ferry compared favorably in quality to those made by private contractors, the weapons produced at Springfield were generally preferred by military authorities. It was generally acknowledged that the Potomac armory excelled at making highly finished pattern and presentation pieces, but could not equal Springfield's record for consistently producing a sound, reliable and—after 1815—a more uniform product" (Smith 1977, p. 101).

guns were "superior to every other kind of small arm now in use." And "the rifle's parts could be mutually exchanged with another thus greatly simplifying the task of making field repairs" (Smith 1977, pp. 201–202). A second military commission found that the machinery developed by Hall to produce his gun was "unparalleled in contemporary practice" (Smith 1977, p. 206).[4]

In 1819 Secretary of War John C. Calhoun arranged to have Hall appointed assistant armorer with responsibilities to undertake the manufacturer of breech-loading rifles at Harpers Ferry. He was to be paid $60 a month and a royalty for use of his machines of $1.00 for each rifle produced. In 1820 he was placed in charge of a separate rifle works at the Harpers Ferry Armory, which enabled him to pursue his machinery development and rifle production with less interference from the armory management.[5]

In spite of the greater sophistication of the production system at the Springfield Armory, guns with functionally interdependent parts were first produced in substantial quantity by Hall at Harpers Ferry. On December 30, 1822, Hall, writing from Harpers Ferry, informed Secretary Calhoun as follows: "I have succeeded in establishing a method of fabricating arms exactly alike & with economy, by the hands of common workmen & in such a manner as to insure perfect observance of any established model" (quoted in Smith 1977, p. 199).

Hall expected that he would have responsibility for producing rifles for both the federal military and the state militias. However, a legal technicality inserted in appropriation legislation by the Congress required that arms for the state militias be produced by private contractors. Confronted with this problem, the U.S. War

4. The role of Eli Whitney in the development of interchangeable parts has been the source of considerable confusion among economic historians. In his classic work *A History of Mechanical Inventions*, Abbott Payson Usher credits Whitney's Connecticut factory with assembling muskets from interchangeable parts shortly after 1800 (1954, pp. 378–380). Subsequent research indicates that Whitney did enter into an agreement with the War Department in 1798 to produce 10,000 muskets by September 30, 1800. Whitney was, however, so heavily involved in litigation over the patent rights on his cotton ginning machine that he neglected to give adequate attention to his gun manufacturing enterprise. The final batch of muskets produced under the contract were not delivered until January 30, 1809. At that time Whitney's factory did not yet have the capacity to produce interchangeable parts for the rifles that he manufactured (Woodbury 1958; Battison 1966).

5. Hall not only developed many of the machines used in the Rifle Works but also the tools with which they were made. Because up to half of the workmen at the Rifle Shop were involved in development, the cost of rifles produced by Hall at Harpers Ferry was slightly higher than at Springfield. Colonel George Bedford, Chief of Ordnance (1821–1842) in the War Department and a strong supporter of Hall's work, was forced to continuously defend Hall's work to the management of the armory, within the War Department, and to the Congress. Bedford considered the Rifle Works an experimental venture but felt compelled to justify the high costs of the rifles produced at Harpers Ferry to the Congress on the basis of the potential savings that would accrue from the adoption of fully mechanized techniques (Smith 1977, pp. 220–228).

Department arranged for Hall to make his technology available to Simon North, an innovative arms maker of Middletown, Connecticut. After some hesitation Hall agreed to provide the necessary technical assistance, and North received a contract with the U.S. Ordnance Department in 1828.[6]

By 1834 North was able to produce rifle components that could be exchanged with rifles made at Harpers Ferry: "For the first time fully interchangeable weapons were being made at two widely separated arms factories" (Smith 1977, p. 212). The system of gun production developed by Hall also influenced arms making at the Springfield Armory. It diffused rapidly to other Connecticut Valley arms manufacturers and elsewhere as employees of the Harpers Ferry and Springfield armories transferred their employment and skills to other gun manufacturers (Smith 1991, pp. 241–250).[7]

In the early and mid-1850s a number of industrial commissions from Great Britain and other European countries traveled to the United States to report on the machine processes used in American manufacturing and to purchase tools and equipment. During a visit to the Springfield Armory, one such committee selected ten muskets, each made in a different year between 1844 and 1853, "which they caused to be taken to pieces in their presence, and the parts placed in a row of boxes, mixed up together." They asked the "workman, whose duty it is to 'assemble' the arms to put them together, which he did—the committee handing him the parts taken at hazard—with the use of a turnscrew only, and as quickly as though they had been English muskets, whose parts had been kept separated" (Rosenberg 1972, p. 91).

Diffusion of the Armory System

As transportation and communication improved and as cheap coal became widely available during the second half of the nineteenth century, "armory practice" slowly

6. Hall was forced to restrict his active management of the Rifle Works beginning in 1837 because of a chronic illness (probably tuberculosis). He died on February 26, 1841, at the age of sixty. His wife was particularly bitter about what she considered the unfair treatment of her husband by the War Department and the management of the Harpers Ferry Armory. After his death she wrote to Colonel George Talbot at the Ordnance Department: "No one but myself can imagine his days of toil and nights of anxiety while inventing and perfecting his machinery. Had he in 1820 listened to the proposals of foreign governments, he might now be enjoying health and prosperity, yet he refused all because he thought by doing so he should benefit his own government" (cited in Smith 1977, p. 218).

7. The "full armory system" was adopted more slowly and less completely in the production of arms for commercial purposes (Howard 1978). For discussion of the role of the Springfield and Harpers Ferry armories in the development of the American machine tool industry, see Rosenberg (1963).

diffused to other branches of manufacturing, usually by the movement of skilled machinists from the New England arms factories to other industries and regions. Assembly of standardized parts became common not only in the manufacturing of firearms, but also in the making of locks, watches, clocks, and sewing machines and in other woodworking and metalworking industries (Hoke 1990; Smith 1991).

The sewing machine industry was the first to adopt armory practice. At the Wheeler and Wilson Manufacturing Company (Bridgeport, Connecticut), the armory system was adapted in 1857 by former employees of the Springfield and Colt armories to the production of sewing machines. The Singer Manufacturing Company did not make the transition to full armory practice until 1873, when it opened a new factory at Elizabethport, New Jersey.

The evolution of the American system was closely associated with the emergence of the machine tool industry (Rosenberg 1963). In 1820 there was no separate identifiable machinery-producing sector. Machinery-producing establishments made their first appearance as adjuncts to factories specializing in the production of final products, especially textiles and firearms. As the capacity of such shops expanded, they began to sell machines, first to other firms in their own industry and then to firms in other industries. With the continued growth in demand for an increasing array of specialized machines, machine tool production emerged by the early 1850s as a separate industry. It played a critical role in the diffusion of machine technology in metalworking industries (Smith 1977, p. 325).

Steady improvements in machine speeds, power transmission, lubrication, gearing mechanisms, precision metal cutting, and many other dimensions of performance were applied in one industrial setting after another. Industries such as textiles, arms, sewing machines, farm machinery, locks, clocks, boots and shoes, and locomotives were unrelated in terms of form of final product, yet very closely related from a technical perspective. Because the specialized machine tool industry dealt with processes common to a number of industries, it became a source of rapid diffusion of machine technology across the whole range of metal-using industries (Nelson and Wright 1992).

The question of why American machine technology had come to occupy an increasingly dominant position by the end of the nineteenth century has been intensely debated by economic historians. Ames and Rosenberg (1968) suggest an induced technical change interpretation. At least part of the explanation lies in relative factor prices—particularly the prices of raw materials (wood and metal) and the wages of highly skilled workers compared to less skilled workers. Ames and Rosenberg also emphasize demand-side factors, such as a relatively stable American arms procurement policy and differences in nonmilitary demand, for example, for inexpensive utilitarian firearms in the United States, as opposed to fine sporting arms in Britain.

During the third quarter of the nineteenth century, the American system, broadly defined as the mass production of precision metal components by a sequence of specialized machines, was adapted to an ever-widening range of products. The development of this new machine technology depended on a high order of mechanical skill, as well as ingenuity in conception and design. Increasingly, the advances were the product of a specialized machine tool industry; they were not the product of institutionalized research and development, nor did they draw in any substantial way on recent advances in scientific knowledge.[8] The advances in machine making and machine using, identified as the American system, set the stage for the emergence of *mass production.*

Mass Production

By the end of the nineteenth century, a number of American industries had achieved high-volume production—what later came to be termed *mass production.* Mass production was made possible by advances in machine technology. Mass marketing was made possible by the development of national rail and telegraph networks and a large domestic market. These industries included new branded and packaged products (cigarettes, canned goods, flour and grain products, beer, dairy products, soaps, and drugs), light machinery (sewing machines, typewriters, cameras, and electrical equipment), and standardized industrial machinery. Although these products were initially developed for the domestic economy, many—including industrial machinery, farm equipment, and other engineering and producer goods—came to dominate international markets (Chandler 1977, pp. 240–296).[9]

These turn-of-the-century achievements have been attributed to the confluence of two technological streams: (1) the continuing advance of mechanical and metal-working skills, and their application to high-volume production of standardized commodities; and (2) the exploration, development, and use of the nation's mineral resource base (Nelson and Wright 1992, p. 1938). Mineral discovery, mineral extraction, and advances in metallurgy drew from, stimulated, and induced some of

8. The period beginning in 1859 was one of remarkable scientific progress. "If one had to choose any fifteen-year period in history on the basis of density of scientific breakthroughs that took place, it would be difficult to find one that exceeded 1859–74" (Mowery and Rosenberg 1989, p. 22). But these advances in science were only loosely related to with advances in technology. "Relatively little of the American performance during this era was based on science, or even on advanced technical education. American technology was practical, shop floor oriented, and built on experience" (Nelson and Wright 1992, p. 1938).

9. In this section I draw substantially on Hounshell (1984, pp. 189–302).

the most advanced engineering developments of the time.[10] The often noted complementarity between capital and natural resources in that era was not merely an exogenous technical relationship, but was also induced by a combination of natural resource abundance and rising industrial wages (Cain and Patterson 1981). This meant that, although American products were often competitive on world markets, the technology employed in their production was often inappropriate to economies with different resource endowment, or to economies in which a mass market had not yet developed (Ruttan 2001, pp. 15–60).

Bicycle Manufacture

The American system of mass production emerged in its most highly developed form at the Ford Motor Company in the first decade of the twentieth century. Early bicycle production, however, represented a transitional technology between the American system that emerged out of New England armory practice and the era of mass production.[11] The bicycle industry was responsible for a number of important technical innovations that set the stage for the automobile industry, including the use of ball bearings and pneumatic tires. The most important innovation, however, was the adoption and development of sheet steel stampings to replace drop forging and machining. In New England armory practice, drop forging and machining were the principal processes used in metal fabrication. Western Wheel Works broke from this tradition by adopting stamping technology to produce frame joints previously imported from Germany. The metal stamping equipment developed by Western Wheel toolmakers enabled it to extend the stamping technology to almost every part of the cycle and to reduce machining to a bare minimum.

The contributions of the bicycle industry to the automobile industry were not

10. The development of the mineral industries represented an example of the contribution of public support for science and technology in the United States. The U.S. Geological Survey under the leadership of Major John Wesley Powell was the most ambitious and successful government science agency of the nineteenth century. Under Powell's leadership the United States achieved world leadership in the training of mining engineers and in mining practice (Nelson and Wright 1992, p. 1938).

11. The manufacture of bicycles in the United States began in 1878 when Albert A. Pope, a Boston merchant who had been importing English high-wheel cycles, contracted with the Weed Sewing Machine Company of Hartford, Connecticut, to manufacture an American version. By the time the safety bicycle was introduced from England in 1887, Pope and several smaller firms had produced in the neighborhood of 250,000 high wheelers. Introduction of the safety bicycle set off a new wave of enthusiasm for the bicycle that reached its peak in the mid-1890s when the industry produced 1.2 million machines. In 1896, production by the Pope firm was exceeded by Western Wheel Works of Chicago (Hounshell 1984, pp. 189–215).

only technical. The bicycle revealed a latent demand on the part of the American public for an effective means of personal transport. It remained for the automobile industry, however, to resolve the problem of assembly that would make possible low-cost mass production of a means of personal transportation.[12]

The stage had also been set for the automobile industry by the remarkable growth of the U.S. economy in the latter half of the nineteenth century. Rapid growth continued through the first three decades of the twentieth century. From 1903, the year in which the Ford Motor Company was organized, to 1926, when the last Model T rolled off the Ford assembly line, net national product grew at a rate of over 7 percent per year—comparable to the rates achieved by the East Asian "miracle countries" from the 1960s into the 1990s. It was this growth in consumer income, combined with the large decline in the real price of the automobile, that made the rapid growth in automobile ownership possible (Hughes 1986, p. 285).

The Model T Idea

Mass production at the Ford Motor Company was a product of Ford's commitment to simplicity in design and efficiency in manufacturing. The transition from production in a poorly equipped job shop to mass production was accomplished by substantial experimentation. Ford himself was a classic mechanic. He had remarkable insight into how machines worked and could be made to work better. He brought together a talented team of young engineers and executives, and encouraged experimentation with fresh ideas for gauging, fixture design, machine tool design, factory layout, quality control, and material handling. Ford production engineers tested and adapted what they found useful from New England armory practice, particularly interchangeable parts, and from *Western practice*, such as pressed steel parts, and added a continuous stream of their own innovations. A first step toward mass production began with eliminating *static assembly* by rearranging machine tools according to the sequence of manufacturing operations rather than by type of machine.[13] A second was the construction of a new factory

12. "The question of who built the first automobile is still a matter of dispute, but the Germans Karl Benz and Gottlieb Daimler were probably the first, with their gasoline-powered vehicle of 1885. Later Armand Peugeot built a workable car in France, and by the 1890's the European auto industry had begun. In the United States, the auto industry dates from September 21, 1893, when the brothers Duryea of Springfield, Massachusetts, who were bicycle mechanics . . . built a carriage driven by a one-cylinder motor. By 1899 about thirty American companies built some 2,500 automobiles for sale" (McCraw 1996, pp. 6–7).

13. Rearrangement of machine tools on the shop floor became possible when tools driven by shafts and belts were replaced by tools driven by electric motors (Devine 1983).

at Highland Park in Michigan, which was designed to facilitate the handling of materials (Biggs 1995).

When Ford made the decision in 1909 to move to Highland Park, he also made a decision that the Ford Motor Company would produce only the Model T with identical chassis for its several variants—runabout, touring car, town car, and delivery car. Workers distributed the necessary parts to each workstation and timed their delivery so that they reached the station just before they were needed. Assembly teams moved from station to station to perform specialized tasks. The first Ford assembly lines for components, such as the magneto coil, were installed in 1913. Within a year virtually every assembly operation at Ford had been put on a moving line basis. By April 1914 the time required per assembly had been reduced to just 1.5 man-hours. The Model T that came off the Ford assembly line represented the ultimate standardized machine. It was small, light, and strong, and contained a minimum of working parts.

In a retrospective article in the *Encyclopaedia Britannica*, Henry Ford (1926, pp. 821–823) articulated the principles of "mass production." To Ford (or his ghostwriter) "mass production" was the method by which "great quantities of a single standardized commodity are manufactured": "Mass production is not merely quantity production for this may be had with none of the requisites of mass production. Nor is it merely machine production, which may also exist without any resemblance to mass production" (Ford 1926, p. 821). According to Ford, the essential principles were (1) the orderly progression of the commodity through the shop, (2) the delivery of parts to the worker, and (3) an analysis of operations into their constituent parts. "Every part must be produced to fit at once into the design for which it is made," he continued. "In mass production there are no fitters" (Ford 1926, p. 822). It is doubtful that the machine tool industry could have met the standard that Ford articulated before 1913 (Hounshell 1984, p. 233).

Perspective

The system of production that the British Commission observed at the Springfield Armory was, as already noted, quite limited before 1840. Initially, only the Army was in a position to subsidize the high cost of moving materials to remote manufacturing locations such as Springfield and Harpers Ferry and to transport large numbers of finished guns to even more remote locations on the western frontier. And only the U.S. War Department could provide the large arms contracts that enabled private manufacturers such as North, Whitney, and Colt to make the large investments necessary to build and equip factories with the machinery to produce the interchangeable parts for gun production.

In his review of the history of the Frankford Arsenal, James J. Farley concludes as follows: "In the first half of the nineteenth century the Ordnance Department, rather than private industry, directed the evolution of the American System of manufacture. It pioneered both uniformity and interchangeability" (1994, p. 48). The emergence of an independent machine tool industry in the United States around the middle of the nineteenth century and of mass production in the first decades of the twentieth were the direct consequences of the investment by the U.S. War Department during the first half of the nineteenth century in the invention of armaments, in the development of machines, and in machine methods of manufacturing.

ACKNOWLEDGMENTS

I am indebted to Nathan Rosenberg and Merritt Roe Smith for comments on an earlier version of this chapter.

REFERENCES

Ames, E., and N. Rosenberg. 1968. "The Enfield Arsenal in Theory and History." *Economic Journal* 78: 827–842.

Battison, E. A. 1966. "Eli Whitney and the Milling Machine." *Smithsonian Journal of History* 1: 9–34.

Biggs, L. 1995. *The Rational Factory*. Baltimore: Johns Hopkins University Press.

Cain, L. P., and D. G. Patterson. 1981. "Factor Bias and Technical Change in Manufacturing: The American System." *Journal of Economic History* 41: 341–360.

Chandler, A. D., Jr. 1997. *The Visible Hand: The Managerial Revolution in American Business*. Cambridge, MA: Harvard University Press.

Devine, W. D., Jr. 1983. "From Shafts to Wires: Historical Perspectives on Electrification." *Journal of Economic History* 43: 347–462.

Farley, J. J. 1994. *Making Arms in the Machine Age: Philadelphia's Frankford Arsenal, 1816–1870*. University Park: Pennsylvania State University Press.

Ford, H. 1926. "Mass Production." In *Encyclopaedia Britannica*, 13th ed., suppl. vol. 2: 821–823.

Hoke, D. R. 1990. *Ingenious Yankees: The Rise of the American System of Manufacturers in the Private Sector*. New York: Columbia University Press.

Hounshell, D. A. 1984. *From the American System to Mass Production, 1800–1932*. Baltimore: Johns Hopkins University Press.

Howard, R. A. 1978. "Interchangeable Parts Reexamined: The Private Sector of the American Arms Industry on the Eve of the Civil War." *Technology and Culture* 19: 633–649.

Hughes, J. 1986. *The Vital Few: The Entrepreneur and American Progress*. New York: Oxford University Press.

McCraw, T. K. 1996. "Henry Ford and Alfred Sloan." In *Management Past and Present: A Casebook on the History of American Business*, ed. A. D. Chandler, Jr., T. K.

McCraw, and R. S. Tedlow, chapter 6. Cincinnati, OH: Southwestern College Publishing, 6–30.

Mokyr, J. 1990. *The Lever of Riches: Technological Creativity and Economic Progress.* New York: Oxford University Press.

Mowery, D. C., and N. Rosenberg. 1989. *Technology and the Pursuit of Economic Growth.* New York: Cambridge University Press.

Nelson, R. R., and G. Wright. 1992. "The Rise and Fall of American Technological Leadership: The Postwar Era in Historical Perspective." *Journal of Economic Literature* 23: 1931–1964.

Rosenberg, N. 1963. "Technological Change in the Machine Tool Industry, 1840–1910." *Journal of Economic History* 23: 414–443.

———. 1972. *Technology and Economic Growth.* New York: Harper & Row.

Ruttan, V. W. 2001. *Technology, Growth, and Development: An Induced Innovation Perspective.* New York: Oxford University Press.

Smith, M. R. 1977. *Harpers Ferry Armory and the New Technology: The Challenge of Change.* Ithaca, NY: Cornell University Press.

———. 1985. "Army Ordinance and the 'American System' of Manufacturing, 1851–1961." In *Military Enterprise and Technological Change: Perspectives on the American Experience,* ed. M. R. Smith. Cambridge, MA: MIT Press, 39–86.

———. 1991. "Review of Ingenious Yankees: The Rise of the American System of Manufacturers in the Private Sector." *Journal of Economic History* 51: 244–246.

Usher, A. P. 1954. *A History of Mechanical Inventions.* Cambridge, MA: Harvard University Press.

Woodbury, R. S. 1958. "The Legend of Eli Whitney and Interchangeable Parts." *Technology and Culture* 1: 235–253.

3

Military and Commercial Aircraft

The U.S. government employed three principal instruments to support the development of a commercial aviation industry: large airmail subsidies, procurement of military aircraft, and support for aeronautics research and development. These efforts were exceptionally successful. Between the late 1920s and the mid-1960s, productivity growth in the air transport industry ran upward of 8 percent per year—more rapid than in any other industry during that period (Mowery and Rosenberg 1982a; Mowery and Rosenberg 1982b; Mowery and Rosenberg 1989).[1]

The aircraft industry was exceptional among manufacturing industries in that a government research organization, the National Advisory Committee for Aeronautics (NACA), was established to support research and technology development (R&D) for the industry (Mowery and Rosenberg 1982b, p. 170). NACA was established in 1915 as an independent agency. Its governance structure included representatives from the several military services and the scientific and engineering communities. NACA adopted the public armory or experiment station model in the development of its aeronautics R&D program (see chapter 2). It was not until NACA was absorbed by the National Aeronautics and Space Administration (NASA) in 1958 that a transition to the private contractor model for public support of aeronautics R&D was initiated.

My primary objective in this chapter is to trace the role of military and defense-related R&D institutions, including NACA and NASA, in supporting technology development and the emergence of a military and commercial aircraft industry

1. In this chapter I draw substantially on Mowery and Rosenberg (1982a, 1982b, 1989), Roland (1985), and Anderson (2002). For a more technical discussion of NACA research, see Gray (1948). For very useful histories of the development of aircraft and the aircraft industry, see Miller and Sawers (1968) and Crouch (2003).

in the United States. A secondary theme is the troubled relationship between scientific research and technology development in the federal programs designed to support the development of the aircraft industry.

Struts, Wires, and Glue

The first successful sustained flight of a heavier-than-air self-powered flying machine was achieved by brothers Orville and Wilbur Wright at Kitty Hawk, North Carolina, on December 17, 1903. The Wright brothers did not invent the airplane; they invented the first *successful* airplane. The initial flight lasted 12 seconds and landed 120 feet from where it started. By 1905 an improved Wright machine had stayed in the air 38 minutes and traveled 24 miles (Anderson 2002, pp. 1–4). In 1908 the Wright brothers spent a triumphal summer in France, during which they made several flights lasting more than 2 hours (Crouch 2003, pp. 103–106). In 1908 the U.S. Army Signal Corps purchased its first airplane from the Wright brothers (Crouch 2003, p. 636).[2]

During the nineteenth century, academic researchers made substantial advances in the understanding of fluid dynamics (Anderson 2002, pp. 42–45). Important contributions were made to aerodynamic theory in the late 1890s and early 1900s by a number of European mathematicians, scientists, and engineers, including Wilhelm Kutta (Germany), Nikolai Joukowski (Russia), Fredrick Lancaster (England), and Ludwig Prandtl (Germany). In the early 1900s Prandtl, drawing on his earlier work on the flow of liquids, articulated the circulation theory of lift. His boundary layer hypothesis is regarded as the most fundamental concept in aerodynamic theory. Prandtl's theoretical work and his research facilities, including laboratories and a wind tunnel, established Gottingen University as the world leader of aerodynamic research (Rosenberg 2001; Anderson 2002, pp. 239–234; Crouch 2003, pp. 124–125).

There had also been substantial effort by craftsmen and engineers to understand the fundamental laws of flight and to build successful flying machines (Anderson 2002, pp. 27–79). But, as in many other areas, there was relatively little cross-fertilization between advances in scientific knowledge and engineering and mechanical practice. The knowledge employed by the Wright brothers and other early

2. "In response to a request from the Ordinance Board of the United States War Department on May 31, 1907, the Wrights submitted a formal offer to sell one of their flying machines for $100,000. The War Department did not accept this proposal. Later that year, however, the United States Signal Corps advertised for competitive bids on an airplane and accepted that of the Wrights . . . for $25,000" (Simonson 1960, pp. 361–362).

aircraft designers drew almost entirely on craft and engineering knowledge and practice. The technology of early flight owed "practically nothing to the relatively mature state of the science of fluid dynamics" (Anderson 2002, p. 45). Significant advances in aircraft design and performance continued to be made by flight enthusiasts, mechanics, and engineers during the decade after the Wright brothers' first flight. Until after World War I, most of the advances in aircraft design after the Wright brothers' demonstrations in France in 1908 were made in Europe.

Growth of the aircraft industry itself was largely demand induced by military procurement during World War I. In Germany and France, and to a lesser extent in Great Britain, Russia, and Italy, military procurement began to expand in anticipation of World War I. In the United States rapid growth was delayed until after the United States entered the war in April 1917. At that time the Curtis "flying boat" was the only product of the U.S. aircraft industry that European governments judged suitable for military service. Curtis began producing its flying boat for the U.S. Navy in 1912 and was already exporting its machines to European countries before the United States entered the war.

When the United States entered World War I, the Allied powers urged the United States to dramatically increase its aircraft production. In 1916 the U.S. industry had produced only 411 airplanes. "The joint Army-Navy Technical Aircraft Board established quotas of 8,075 training aircraft; 12,000 frontline pursuit, bombing and observation aircraft; and 41,000 engines" (Crouch 2003, p. 188). An Aircraft Production Board was established and assigned responsibility for organizing production and distributing contracts. It bought land and built factories for its contractors, hired workers, subcontracted parts production, and supervised production.[3]

The European allies advised the United States to produce proven European aircraft types rather than develop untried American designs. In the case of aircraft engines, however, a decision was made to use the recently developed 410-horsepower Liberty 12-A engine for the largest planes. In spite of technical difficulties and accusations of corruption, the U.S. aircraft industry, then composed of approximately three hundred firms and employing some 175,000 workers, "produced 12,894 aircraft and 41,983 engines between April 1917 and November 1918" (Crouch 2003, p. 192).

By the end of the war, the military airplane had evolved into an increasingly

3. During World War I the Navy operated a Naval Aircraft Factory located in Philadelphia. The factory built a number of "flying boats" during the last year of the war. After the war the factory became embroiled in controversy about aircraft production by the Navy during the period of postwar depressed demand. An attempt was made to deflect criticism by converting the factory to aeronautical research devoted to the special problems of naval aircraft (Trimble 1986).

effective flying machine. It achieved its greater speed and improved performance primarily because of the brute force of its increasingly more powerful engines.[4] It did not, however, incorporate any dramatic revolutionary breakthroughs in design. The aircraft employed by both the German and Allied armies in World War I were evolutionary descendants of the first generation of Wright Flyers: "For the most part they were 'souped up' Wright Flyers" (Anderson 2002, p. 152).

The one revolutionary aircraft introduced during World War I was the all-metal Junkers J-1 developed by Hugo Junkers. It was a monoplane and had no external struts or wires. Its first flight took place on December 11, 1915. "It was designed, built, and flown in an era when every other airplane was vegetable airplane—made from wood, fabric and glue" (Anderson 2002, p. 171). The J-1 was followed by the all-metal Junkers J-13 commercial transport in 1919. But it was not until the introduction of the Ford Trimotor in the mid-1920s that all-metal aircraft began to find a secure place in the airline industry.

The momentum that had developed in the U.S. aircraft industry during World War I was quickly dissipated in the years immediately after the war. Aircraft production in the United States declined from 14,020 (13,991 military) in 1918, to 328 (256 military) in 1920 (U.S. Bureau of the Census 1960, table Q345–351). Only intervention by the Army and the Post Office to develop a national postal airmail service prevented the industry from complete collapse (box 3.1). It was not until 1926 when the Congress passed the Navy Five Year Program and the Army Five Year Procurement Programs that substantial recovery from the post–World War I aircraft market depression was assured.

The NACA Era

Well before the beginning of World War I, there was substantial concern within the aviation community that the United States was lagging behind Germany, France, the United Kingdom, and Russia in institutionalizing aircraft R&D capacity. During the late nineteenth century, a number of European war ministries had created aeronautical research, development, and design facilities (Crouch 2003, p. 135). During the years 1908 to 1913, the United States ranked fourteenth, just below Brazil and above Denmark, in government expenditures on aircraft development (table 3.1).

4. The German Albatros DVa and the French SPAD XIII fighter planes were examples of the best military design as of 1917. The 1903 Wright Flyer had a 12-horsepower engine and a speed of 30 miles per hour. The Albatros DVa had a 180-horsepower Mercedes engine and a top speed of 116 miles per hour. The SPAD XIII had a 200-horsepower Hispano-Suiza engine and a top speed of 120 miles per hour (Anderson 2002, pp. 131–132).

Box 3.1. Postal Subsidies for Airline Development

For almost two decades the Post Office Department played an active role in airline and aircraft development. In 1918, using aircraft and pilots borrowed from the Army, the Post Office opened airmail service between New York and Washington, DC.

A successful public postal air service was first established among major cities along the East Coast, with surplus World War I airplanes and modified de Havilland bombers (DH-4) powered by Liberty engines. By September 1930 the last link in a transcontinental airmail service had been established. By 1925 installation of rotating beacons at airports across the country made night flight feasible (Von der Linden 1991, pp. 5–6; Von der Linden 2002, pp. 1–34).

In the mid-1920s Congress passed several pieces of legislation that changed the structure of the airmail system. The Contract Mail Act (the Kelly Act) of 1925 was designed to replace the Post Office in the transport of mail with contract carriers. Routes were to be awarded by competitive bidding. The Air Commerce Act of 1926 was an organic act that consolidated previous legislation governing air carrier operations. In addition, it authorized the U.S. Commerce Department to designate national airways, license pilots and aircraft, investigate accidents, and promote research and development of aerial navigation aids. A 1926 amendment to the Kelly Act authorized payment based on weight for contract postal air services. The effect of this legislation was to induce the formation of numerous potential carriers hoping to take advantage of profitable mail contracts (Von der Linden 1991, pp. 6–8). These legislative acts were implicitly designed to subsidize the development of airlines. By 1927 all airmail was handled by contract carriers (Holley 1964, p. 12).

In 1930 the Kelly Act was again amended, by the McNary-Watres Act, to change the formula by which airlines were compensated for transporting airmail. It replaced the compensation-by-weight formula with a payment by available volume. A price of $1.25 per cubic feet per mile would be paid to contractors "regardless of whether the space in the aircraft was filled" (Von der Linden 1991, p. 21). This change in payment schedule encouraged airlines to squeeze a few passengers into the unused space. It also encouraged aircraft manufacturers to develop and place larger aircraft in the airmail service, including the Ford Trimotor, the Curtis Condor, and the Douglas DC-2 (Holley 1964, pp. 14–15).

Herbert Hoover's new Postmaster General, Walter Folger Brown, moved rapidly to take advantage of the authority granted to him under the McNary-Watres Act to attempt to rationalize the rapidly growing airline industry. Brown used his power to grant contracts to force the merger of smaller airlines into larger national or regional carriers such as United and Transcontinental and Western (TWA; Von der Linden 1991, p. 20; Von der Linden 2002, pp. 85–105). These steps were taken with the conscious effort to develop a sophisticated passenger aircraft industry in support of a national air transport

(continued)

Box 3.1. *(continued)*

system. By 1933 four major airlines (American, TWA, Eastern, and United) linked the major U.S. cities. This airline structure remained dominant until well into the post–World War II era.

In February 1934 President Franklin D. Roosevelt, troubled by the appearance of corruption in the awarding of airmail contracts and the size of the subsidies involved, issued an executive order canceling the all airmail contracts and transferring operations to the U.S. Army. "When the Army attempted to fly the mail on short notice, lacking adequate equipment and training for the task, they were beset with disaster. After a week of midwinter flying and almost daily crashes, the sum of catastrophes stood at five pilots dead and six injured" (Holley 1964, p. 15).

Soon afterward the president rescinded his order and began negotiation to return the mails to private carriers. Congress passed two airmail acts in 1934 that abandoned the subsidies implicit in previous airmail legislation and returned to an emphasis on low bids. Legislation prohibiting the holding companies that had been organized in the 1920s to consolidate airlines, aircraft manufacture, and engine production under single management were barred from bidding on airmail contracts. The effect was to bring about dissolution of the trusts (Von der Linden 2002).

Sources: Holley (1964, pp. 11–20); Von der Linden (1991, pp. 5–9); Von der Linden (2002).

The Aeronautical Society initiated efforts in 1911 to obtain political support for the establishment of a national aeronautical research laboratory to be administered by the Smithsonian Institution and located at the National Bureau of Standards. It was not until 1915, however, that concern about the imminence of U.S. participation in World War I precipitated sufficient convergence of scientific, military, commercial, and political interests to mobilize a successful effort to pass legislation, as part of a naval appropriation bill, to establish an National Advisory Committee for Aeronautics—later the National Committee on Aeronautics (NACA; Roland 1985, vol. 1, pp. 1–25).[5]

5. The organic legislation that established NACA specified a 14-member governing committee consisting of 7 government representatives—2 each from the War and Navy Departments and one each from the Weather Bureau, the Bureau of Standards, and the Smithsonian Institution—

Table 3.1. Estimated Government Expenditures on Aviation, 1908–1913

Country	Expenditures (in 1913 U.S. Dollars)[a]
Germany	$28,000,000
France	$22,000,000
Russia	$12,000,000
Italy	$8,000,000
Austria	$5,000,000
England	$3,000,000
Belgium	$3,000,000
Japan	$1,500,000
Chile	$700,000
Greece	$660,000
Bulgaria	$600,000
Spain	$550,000
Brazil	$500,000
United States	$435,000
Denmark	$300,000

Source: Crouch (2003, pp. 134–135).

[a] In addition to official appropriations, several leading aeronautical powers also established public subscriptions that provided additional support to their aeronautical industries: Germany, $3,500,000; France, $2,500,000; Italy, $1,000,000; and Russia, $100,000.

Policy Advice

During World War I, NACA served primarily as an advisory and consultative board. Its initial effort was the sponsorship of a national survey of aeronautical research and development in universities and in the aircraft industry. The National Research Council (NRC) was, however, the primary instrument employed by the military in sponsoring aeronautics research and development directly relevant to the war effort during World War I. NACA did play a constructive role in resolving

and no more than 7 private members. Members were appointed by the president and served without compensation. The chairman would be elected by the members. At the first meeting of the committee on April 23, 1915, it was agreed that a 7-member subcommittee of the main committee, designated the Executive Committee, should be in charge of administrative matters. At the first meeting Charles D. Walcott, secretary of the Smithsonian Institution, was elected chairman (in absentia).

disputes over patent rights between the Wright and Curtis interests and over aircraft engine design, disputes that threatened to halt aircraft production.[6] It was also able to obtain congressional approval for its efforts to initiate development of research facilities near the mouth of the Black River near Hampton, Virginia.[7]

During the early 1920s, NACA became involved in a series of policy disputes over the organization and management of civil and military aviation by the federal government. These issues were resolved by the Air Commerce Act of 1926, which rejected the establishment of a unified military air force, established a Bureau of Aeronautics in the U.S. Department of Commerce, and authorized the secretary of commerce to encourage and regulate civil and commercial aviation. With the passage of the 1926 act, NACA backed away from the policy role implied by its original authorization, in order to focus on "the scientific study of flight with a view to their practical solution" (Roland 1985, vol. 1, pp. 51–71).

Wings and Propellers

From the end of World War I until well into the 1920s, NACA was confronted with the development of research facilities and staff at the then relatively isolated Langley Field, and with an uncooperative Army Air Service administration.

During its first decade the two most significant contributions of the NACA research program to aeronautical knowledge and technology development involved the analysis of propeller design and the construction and use of advanced wind tunnels. The initial propeller design analysis and tests were carried out under contract with William F. Durand, professor of mechanical engineering at Stanford University. Durand and his associate, Everett P. Lesley, focused their efforts on the effects of propeller design on propulsion efficiency. Initial tests, using data generated from the operation of a small wind tunnel that Durand and Lesley constructed on the Stanford campus, involved the performance of variations in standard propeller blade forms. As their research proceeded, they employed the method of *parameter variation* to test the full range of potential propeller designs.[8]

6. The Wright patent suits were resolved in spring 1917 by the creation of a government-approved patent pool, the Manufacturers Aircraft Association. In April 1917 the Aircraft Production Board was created to establish priorities and arrange contracts for the production of aircraft and engines (Crouch 2003, p. 191).

7. The location at what later became Langley Field involved a joint decision by the Army Air Service and NACA. A major attraction for NACA was that the Army planned to establish an aviation experiment station and proving ground at Langley Field. It was a major disappointment to NACA when the Army decided instead to locate its experimental activities at McCook Field near Dayton, Ohio (Roland 1985, vol. 1, p. 80).

8. The method of *parameter variation* involves "the procedure of repeatedly determining the performance of some material process or device while systematically varying the parameters that

The results of their work from 1916 to 1926 had little impact on propeller design until the 1930s, when the engineering problem of changing propeller pitch in flight was resolved, which made the variable pitch propeller technically feasible. The data produced by the Durand-Lesley studies did, however, enable airplane designers to improve their ability to match the design of the propeller with engine and airframe design (Vincenti 1990, p. 158).

As the work by Durand and Lesley was being completed, NACA was involved in the design and construction of a variable density wind tunnel at Langley Field, under the direction of Max Munk, its brilliant but erratic chief scientist, who had received his training with Prandtl at Gottingen. Munk's work "provided a new and illuminating way to think about airfoils and resulted in a basic shift in airfoil design" (Roland 1985, vol. 1, p. 36). His state-of-the-art wind tunnel enabled NACA to test airfoil design parameters by varying conditions to simulate those encountered in flight. Munk also built a wind tunnel designed specifically for research on propellers, before he resigned as a result of a bitter personality clash with NACA administration. Monk's work on airfoils was the first NACA research to make a major impact on aircraft design. By designing, modeling, and testing a whole series of airfoils in which characteristics were slightly and systematically varied, NACA was able to provide designers with wing section designs for every purpose (Roland 1985, vol. 1, pp. 92–96). In 1931 a full-scale wind tunnel was constructed that was able to test the performance of an entire aircraft by simulating flight under controlled conditions (Gorn 2001, p. 98).

The appointment of board member Joseph Ames of Johns Hopkins University as NACA chairman in 1927 ushered in a period of rapid growth in the agency's facilities and staff. The success of its research program was followed by rapid growth of NACA's budget and staff, and diversification of its research portfolio, including such esoteric areas as man-machine interaction in exploring the tradeoff between stability and control in achieving improvements in flying qualities for pilots (Vincenti 1990).[9]

From its beginning NACA had adopted a policy of being especially responsive to requests for assistance from the military. Under the Ames chairmanship NACA became even more responsive to concerns of both the military and civil aircraft

define the object of interest or its condition of operation" (Vincenti 1990, p. 139). For a discussion of the development and role of parameter variation in advancing engineering knowledge see Vincenti (1990, pp. 159–169).

9. The Air Force also conducted an in-house research program. The variable-pitch propeller was developed by Frank Caldwell, chief engineer of the propeller department at the Army Air Force aeronautical research and development center at McCook Field. The variable-pitch propeller allowed the propeller pitch to be changed from the cockpit to achieve maximum efficiency. Caldwell received the 1929 Collier award for development of the variable-pitch propeller (Anderson 2002, pp. 253–280).

industries. An important example was what came to be known as the *NACA Cowling.* In the mid-1920s both the military services and industry representatives urged NACA to investigate the design of air-cooled radial engine cowlings.[10] Wind tunnel tests were begun in July 1927, and by the end of the year designs and blueprints were circulated to industry representatives. Roland (1985) notes, "The 60-percent reduction in drag and a 14-percent increase in speed predicted by the NACA were demonstrated in February 1929 when a Lockheed Air Express equipped with the NACA Cowling established a transcontinental speed record of 18 hours and 13 minutes" (vol. 1, p. 116). The aircraft industry quickly adopted the NACA Cowling.

The research approach involved in the design of the NACA Cowling, like that employed in efficient propeller design, involved extensive use of experimental parameter variation (Anderson 2002, p. 219). The development of the NACA Cowling had been entirely empirical. It was not until the late 1930s that Theodore Theodorson, then head of the Langley Physical Research Division, presented a scientific interpretation of the aerodynamic efficiency of the NACA Cowling at the sixth annual meeting of the Institute of Aeronautical Sciences (Anderson 2002, pp. 218–228).

In the early 1930s the NACA Langley Field facility was recognized as the leading aeronautics research center in the world. By the mid-1930s, however, there was growing concern in both the military and commercial aircraft industries that the budget stringency under which NACA was being forced to operate was again resulting in a substantial lag in the quality of its research compared to that in Europe, particularly in Germany. These concerns led to a decision to establish a new research center at Moffett Field, near Sunnyvale, California. The location was selected to be near the West Coast airframe manufacturing industry. By 1940 a decision had been made to develop a NACA aircraft engine research laboratory near Cleveland, Ohio. The location was selected to be near the aircraft engine manufacturing industry in the Midwest (Roland 1985, vol. 1, pp. 155–166). A third new facility, the Dryden Flight Research Center, was established at Edwards, California, in 1944.[11]

10. In the mid-1920s "airplanes could be divided into two general categories on the basis of the type of piston engine used—the liquid-cooled in line engine or the air-cooled radial engine. Air-cooled radial engines had several advantages for aircraft design: lower weight per horsepower, fewer moving parts and lower maintenance costs" (Anderson 2002, p. 218).

11. When NACA began to develop its research program, it initially elected to limit its effort in aircraft engine research. The Bureau of Standards was already engaged in aircraft engine research. It was not until 1940, when NACA decided to establish an engine research laboratory near Cleveland, that it committed itself to a major aircraft engine research program. Its research program was initially directed to improvement in the performance of piston airplane engines

Most of the advances in design and performance that resulted from the NACA program before World War II were "dual use"—applicable to both military and commercial aircraft. Military support for new aircraft development also became a source of technical skills, knowledge, and innovation directly relevant to the manufacture of commercial aircraft (Hunsaker 1952; Mowery and Rosenberg 1982b; Gorn 2001). The technical advances based on NACA research were achieved at a remarkably low cost. The "total appropriations for NACA research between 1915 and 1940 approximated $25 million" (Mowery and Rosenberg 1982b, p. 170).

On the eve of World War II, the United States was building the world's best commercial airliners and had developed the world's largest commercial airline system (Constant 1980, p. 151). The Douglas DC-3 became the most successful commercial airplane of its time. It was also the most technologically advanced.[12] When the DC-3 production line was shut down at the end of World War II, 10,926 had been produced—10,123 for the military and 803 for commercial airlines. Surplus military DC-3 planes were sold to commercial airlines. The DC-3 remained a workhorse for commercial airlines around the world until well into the 1960s (Anderson 2002, pp. 195–201).

The aircraft (and airline) industries were, however, much less successful in attempts to achieve a stable economic structure than in realizing a mature technology. Many of the numerous firms in the industry were initially established or financed by wealthy hobbyists, designers, and entrepreneurs "whose enthusiasm for aircraft defied rational business calculation" (Vander Meulen 1992, p. 57). Revenues were highly variable because of dependence on unstable military demand (Simonson 1960). The numerous independent firms were still not in position to conduct substantial independent research in aeronautics and aircraft design.

Military and defense-related support clearly played a major role in the development of the pre–World War II aircraft industry. Military considerations were a primary motivation for the establishment of NACA and the support of NACA's

(Roland 1985). For a history of the development of the aircraft piston engine, see Heron (1961). Heron notes that the development of the piston engine was almost entirely empirical, "a product of test bed running and the process of break, burn and melt" (1961, p. 115).

12. The DC-3, introduced in 1935, embodied no new revolutionary design features or new technology. Each aspect of the design revolution, such as the NACA Cowling, wing flaps, retractable landing gear and others, had been tested and demonstrated separately (Miller and Sawers 1968, pp. 98–127). "But what was revolutionary about the DC-3 was that it contained in its design, for the first time, all of the features of a mature propeller-driven airplane and that the designers of the DC-3 combined all of these features in a synergistic fashion in one of the most technologically successful airplanes in history" (Anderson 2002, p. 201). For more than thirty years, commercial airliners got bigger, faster, more powerful, and able to fly longer distances, but they remained evolutionary from the DC-3 (Miller and Sawers 1968, p. 37).

R&D. Demand for military aircraft provided an important inducement for advances in airplane design by the aircraft industry. It remains an open question whether in the absence of military-related R&D and aircraft demand a mature propeller-driven commercial aircraft, epitomized by the DC-3, would have become available to the airline industry in the 1930s (Mowery 2004). The DC series of planes was initially developed for commercial rather than military use. Their development was financed by Douglas. But they incorporated technologies that had been developed in the production of military aircraft and aircraft engines extending back to World War I.

Jet Propulsion

The first three decades of the twentieth century experienced very substantial progress in aerodynamic theory. These advances, primarily European in origin, had little impact, however, on aircraft design and performance: "The design revolution that brought about the era of mature propeller-driven aircraft drew primarily on empirical data, innovative thinking and hard experience" (Anderson 2002, p. 239). On the other hand, the fact that airplanes were flying higher and faster was a stimulus to academic researchers to investigate the physical laws that governed the performance of airplanes. By the mid-1930s the annual industry conferences hosted by NACA were beginning to reflect a narrowing of the gap between advances in aerodynamic theory and design practice (Anderson 2002, pp. 239–245).[13]

These advances forced recognition, on the part of both the engineering and scientific research communities, of the physical constraints on further development of the piston-propeller system of aircraft propulsion.[14] Further advances in drag reduction, propeller efficiency, weight reduction, and high-altitude flight could be expected to contribute only marginally to performance (Anderson 2002, pp. 276, 280). What had appeared, since the time of the Wright brothers, to be a "path-

13. See, for example, the discussion of NACA research on compressibility. NACA was clearly the world's leading research institution in the area of compressibility effects by the mid-1930s (Anderson 2002, pp. 298–308).

14. In propeller-driven aircraft, air is pushed backward by the propeller, thus forcing the plane forward. In jet engines, air gathered from the atmosphere is compressed, mixed with fuel and burned, passed through a turbine, and exhausted in a powerful jet of hot gasses. The three types of gas turbine propulsion systems are (1) the *turboprop*, which uses an internal combustion gas turbine to drive a conventional propeller; (2) the *turbojet*, which uses an internal gas turbine as a gas generator and a reaction propulsion nozzle as a thrust producer; and (3) the *turbofan*, which has a turbojet at its core, but some of the power from the gas generator is used to turn a large ducted fan that acts much like a propeller. By the late 1990s almost all commercial jet airplanes and many military jets were powered by turbofan engines (Roland 1985, vol. 1, p. 187; Bonaccorsi and Giuri 2000, p. 853; Anderson 2002, pp. 336–338).

dependent" trajectory of aircraft propulsion development was approaching a dead end (for more on the concept of path-dependent technical development, see chapter 1).

It was not immediately apparent to American aeronautical engineers and scientists that the transition to a new technical trajectory would depend on the development and adoption of the jet engine.[15] The application of the jet engine for flight would require advances in aerodynamic theory and design principles that were not yet available to the American aircraft and airline industries. However, investigation of the aerodynamics of high speed and of the reliability of turbojet components had already convinced a small group of innovative engineers, led by Frank Whittle in the United Kingdom and Hans von Ohain in Germany, that a transition to a turbojet system of propulsion would enable aircraft to achieve higher levels of performance—to fly at a higher altitude and at greater speed (St. Peter 1999, pp. 3–58).

Both Whittle and von Ohain initiated their jet engine development work with their own resources. After substantial delay Whittle received support from the British Air Ministry.[16] In Germany the potential importance of the prototype jet engine developed by von Ohain was immediately recognized by the aircraft manufacturer Ernst Heinkel. By mid-1937 both the Whittle and von Ohain engines had been operated successfully on test stands. The first successful flight of a jet-powered aircraft, the Heinkel He 178, took place on August 27, 1939. The first successful flight by a British aircraft, the Gloster E-28, took place on May 15, 1941. Both planes were designed as prototype military aircraft (Constant 1980,

15. As early as 1923 NACA had contracted with the National Bureau of Standards to conduct an assessment of the potential application of jet propulsion. The report indicated that fuel consumption and maintenance costs would be substantially higher for jet aircraft than for piston engines with propellers: "The conclusions were correct for the 250 miles per hour or less flight speeds considered in the report. However they were not applicable to high speed flight near or above the speed of sound" (Anderson 2002, p. 285).

16. While serving at the Royal Air Force (RAF) Staff College in the late 1920s, Whittle developed a proposal for jet aircraft development. Both the Air Ministry and the aircraft industry found Whittle's proposal too radical. In 1935, with financial assistance of two ex-RAF officers, he formed a private firm, Power Jets, to exploit his ideas. With the assistance of the Thompson-Huston Company, Whittle had an engine ready for testing in February 1937. Although the test was only partly successful, the Air Ministry made a small grant to Power Jets to continue the tests. After successful tests in October 1938, Whittle received a contract from the Air Ministry to build a flight engine. In April 1940 the Air Ministry selected the Rover Company as engine manufacturer and Gloster Aircraft to build the airframe for a jet fighter. The Gloster plane, powered by a Whittle engine, made its first flight in May 1941. It was not until 1944, however, that production problems were worked out and the Gloster Meteor entered squadron service. The initial lack of enthusiasm on the part of the Air Ministry was due to reluctance to divert funds from efforts to improve the performance of the piston-propeller engine (Cook 1991, pp. 97–114; Bobo 2001).

pp. 178–207; Anderson 2002, pp. 285–289). In retrospect, it is clear that the failure of NACA to correctly assess the potential of jet aircraft technology in the late 1930s and early 1940s must be regarded as a critical lack of foresight on the part of the NACA advisory committee and management.

The major emphasis was on mass production of existing designs (Holley 1964, pp. 304–310). Virtually every U.S. airplane that saw service during World War II was designed before the war. It would not be until well into the first postwar decade that the scientific and design problems confronting the development of military and commercial jet aircraft would be resolved. The failure to develop jet aircraft research and development capacity in the late 1930s left NACA politically vulnerable and in a weak position to obtain the resources needed to reclaim its role as the U.S. lead institution in aircraft design and technology after the war.

Even as Germany and Britain were pursuing advanced jet aircraft research and development in the late 1930s, the NACA staff and board remained skeptical of the technical and economic viability of gas turbines for aircraft propulsion (Hunsaker 1952). In early 1941 Army Air Force General Hap Arnold, during a visit to Britain, became aware of British and German progress in jet engine development. On his return to the United States, he wrote to Vannevar Bush, then NACA chairman, emphasizing the importance and urgency of jet propulsion. Later that year a NACA Special Committee on Jet Propulsion was formed. The Special Committee urged the Army Air Force to contract with General Electric, Westinghouse, and Allis Chalmers, which had recently initiated jet engine development programs, to intensify its efforts to develop gas turbines for aircraft propulsion (Roland 1985, vol. 1, pp. 186–193).

The production of the first jet engines in the United States during World War II drew on British technical assistance and was financed by the Army Air Force. The Army Air Force arranged with the British to acquire a Whittle jet engine, and several British engineers, including Captain Whittle, were brought to the United States to provide technical assistance (Gray 1948, p. 279). In September 1941 the Army Air Force decided to put an aircraft powered by a Whittle jet propulsion engine into production. General Electric would build the engine, and the airplane would be produced by Bell Aircraft Corporation.

NACA was not informed of the Air Force initiative. It was advised that rather than becoming involved in jet engine development, it should concentrate on better engines for fighter planes. In 1943, however, the Air Force arranged for NACA engineers to initiate a research program to enhance the performance of the Whittle engine (Gray 1948, pp. 275–280). It required more than a decade of learning by doing, and using by manufacturers and the military, before jet engines could be maintained and operated with sufficient reliability and efficiency to begin to win a secure place in U.S. commercial airline fleets (Constant 1980; Heppenheimer 1995).

Demonstration of the vulnerability of American heavy bombers to Russian jet fighter planes early in the Korean War was decisive in motivating the transition from the piston-propeller system by the U.S. military services. Boeing Aircraft and Transport Company's experimental jet-propelled bomber, the XB-47, had emerged out of a period of intense interaction between Boeing designers and the U.S. Army Air Forces Project Office during the mid-1940s. Anderson has characterized the XB-47 as an early example of the second design revolution. The swept wings and the engine pods slung below the wings were radical new technological features.[17] They enabled the Boeing engineers to achieve a dramatic leap in performance. This new configuration established a new technological trajectory for both military and commercial jet airplanes (Anderson 2002, pp. 244–347). "The design of the B-47 had been revolutionary. The design of the 707 was evolutionary from the B-47" (Anderson 2002, pp. 347–348).

The world's first commercial jet airliner, the British-built Vickers Viscount turboprop powered by four Rolls Royce Dart engines, made a first experimental flight in 1948. The de Havilland Comet began scheduled service between London and Johannesburg in 1952. Three of the de Havilland Comets crashed in 1953–1954, because designers did not understand that repeated high-altitude pressurization would result in metal fatigue and cause catastrophic failure of the aircraft's fuselage.[18]

Several explanations have been advanced for the failure of NACA to make an early commitment to jet aircraft development. One, as noted earlier, was simple

17. The concept of a swept-wing design for high-speed flight was first advanced by a young German aerodynamicist, Adolf Busemann, at an international conference in Italy in 1935. The military significance of the paper was immediately recognized and classified by German Luftwaffe officials. On May 7, 1945, a U.S. technical team led by Theodore von Karman arrived at research laboratories at Pennemunde and Braunschweig, and appropriated a large body of very advanced aerodynamic technical material. They also arranged, under Operation Paperclip, for Busemann to be assigned to the NACA laboratory at Langley. The concept of swept-wing design for high-speed flight had been independently developed by Robert T. Jones at Langley in 1945. Anderson argues that it is unlikely that the swept wing would have revolutionized airplane design so soon after the war if it had not been for Jones's independent discovery of its advantages (Anderson 2002, pp. 322–327).

18. For a more detailed review of the post–World War I development of commercial aircraft development and civil aviation, see Cook (1991) and Higham (2003). It was not until well into the 1950s that it became clear that jet aircraft were the most economical means of long-distance air transportation. For discussion of the sources of efficiency gains in jet aircraft from the late 1940s through the late 1950s, see Sahal (1981). For greater technical detail with primary emphasis on development of jet engines to meet military specifications, see St. Peter (1999, pp. 329–341).

path dependence—dominance in one area of technology tends to obscure opportunities in related areas (Ruttan 2001, pp. 112–116). NASA historian Alex Roland characterizes propeller-driven path dependence at NACA more pungently: "Give an (aeronautical) engineer a wind tunnel and he will use it—and it will use him. The NACA engineers at Langley Field, possessed of the best research equipment in the world, climbed into their tunnels and promptly lost sight of events outside those narrow chambers" (1985, vol. 1, p. 108). Mowery and Rosenberg are more generous. They attribute NACA's conservative approach to lack of a sense of urgency on the part of NACA's prewar military constituency and to lack of interest in the development of radical new engine technology on the part of the U.S. commercial airline industry until well into the 1950s (1982b, pp. 170–171).

NACA made substantial contributions to advances in aircraft technology immediately before and during World War II. NACA's research on deicing led to the development of thermal deicing through heat exchange with exhaust gases. The low-drag laminar flow wing design of the P-51 fighter aircraft was an important NACA wartime contribution. NACA also gave major attention to the refinement and testing of prototype models of military aircraft and to consulting with aircraft manufacturers on technical problems of aircraft production. Every American airplane and every aircraft engine that was deployed in World War II had been tested and improved by NACA engineers. However, it was widely asserted by both NACA senior staff and by knowledgeable observers in the aircraft industry that these and many other wartime contributions came at the expense of more fundamental aeronautical research (Roland 1985, vol. 1, pp. 173–198; box 3.2).

High-Speed Flight

NACA entered the post–World War period with staff, facilities, and funding that had quadrupled since the early 1940s (Roland 1985, vol. 1, p. 60). It had what it interpreted as a mandate to be the central government institution for aeronautical research.[19] It hoped to make a transition from its "quick fix" role during World

19. In 1945 NACA presented to the Senate Special Committee Investigating the National Defense Program, chaired by Senator James M. Meade, a draft of a National Aeronautical Research Policy that had been worked out in consultation with industry and the military services, a policy by which NACA would be assigned responsibility for aeronautical research, industry would assume responsibility for design and development, and the armed services would assume responsibility for testing and evaluation. In March 1946 a slightly revised version of the draft policy was endorsed by the Army, the Navy, the Civil Aeronautics Administration, and the newly established NACA Industry Consulting Committee (Hunsaker 1952, p. 24; Roland 1985, vol. 1, pp. 203–207).

Box 3.2. Fundamental Research at NACA

The organic law establishing the National Advisory Committee for Aeronautics (NACA) provided: "That it shall be the duty of the Advisory Committee for Aeronautics to supervise and direct the study of the problems of flight, with a view to their practical solution."[a] This charge has throughout NACA history (1915–1958) been a source of considerable confusion and tension. NACA historian Alex Roland insists that within its first decade NACA turned "to an engineering orientation that it never thereafter abandoned" (1985, vol. 1, p. xiv).

In the 1920s efforts were made to complement the NACA engineering orientation by strengthening academic research in aeronautics. The Daniel Guggenheim Fund for the Promotion of Aeronautics made a number of endowment grants to aeronautical engineering programs at the California Institute of Technology (Caltech), the Georgia School (later Institute) of Technology, the Massachusetts Institute of Technology, New York University, Stanford University, the University of Michigan, and the University of Washington (Hallion 1997). At the time Caltech and Stanford University received the Guggenheim grants, aeronautics research at these institutions already had a strong scientific orientation. Between 1916 and 1926, Stanford University mechanical engineering professors W. F. Durand and E. P. Lesley applied the method of parameter variation to design wind tunnel experiments to evaluate propeller performance (Vincenti 1990, pp. 137–169). In 1930 Caltech was able to attract Theodore von Karman, one of Prandtl's most distinguished students at Gottingen, to its faculty. Von Karman played a leading role in Caltech's subsequent success in establishing itself as a leading academic center for aeronautical science and engineering (Hanley 1982; Rosenberg 2001).[b]

The tension between scientific and engineering orientation at NACA was reflected in debates about post–World War II research policy. In 1944 Edwin Hartman, NACA western research coordinator, argued that after the war the committee should direct most of its attention to development. He insisted that research for which the committee had made the largest contributions and received the most credit was in development. Hartman cited the development of the NACA Cowling and the research on deicing as examples (Roland 1985, vol. 1, pp. 234–237). Roland also notes that at the end of World War II one of the most urgent challenges facing NACA leadership was to respond to the criticism that NACA's wartime neglect of fundamental research had resulted in a "lamentable shortage of fundamental data" (Roland 1985, vol. 1, p. 234).

The problem was not that NACA engineers and scientists did not conduct important fundamental or scientific research. Rather, it was that fundamental or scientific research at NACA was induced by response to specific engineering problems rather than in response to an externally directed imperative to advance scientific frontiers. Rosenberg (2001) has stated the issue that NACA confronted with particular clarity: "Aeronautical engineering makes extensive use of sources of information that do not draw on fundamental science because the specificity of aircraft designs requires information that cannot be deduced

(continued)

Box 3.2. (*continued*)

from the principles of aerodynamics" (p. 10). The critics, blinded by the linear model of the relationship between basic research and technology development, were trying to push NACA in a direction that was not consistent with aircraft technology development.

Between 1929–1954 the NACA staff received five Collier Trophy research awards: (1) for developing the NACA Cowling, which greatly reduced the drag of radial air-cooled engines (1929); (2) for developing thermal deicing through heat exchange with exhaust gases (1946); (3) for developing a prototype research aircraft (the X-1) that broke the "sound barrier" (1947); (4) for the development of a transonic wind tunnel that could produce reliable data on the most troublesome speed ranges (1951); and (5) for discovery and verification of the area rule of aeronautic flow, which established the optimal relationship between the girth of a fuselage and the wings in transonic flight (1954). Of these, only the last involved fundamental scientific research (Roland 1985, vol. 1, pp. 114, 235, 250, 256, 280).

My own reading is that the inconclusive nature of much of this discussion was based on a dialogue constrained by an intellectual commitment to the traditional linear model of the relationship between advances in science and advances in technology on the part of NACA staff and administration, and by NACA critics both within and outside the agency (figure 3.1, top panel). The NACA research program could more accurately be interpreted by an interactive model in which scientific and technical knowledge is drawn from a common pool, interacts and is extended during the research and technology development process, and feeds back into the common pool (figure 3.1, lower panel).

Roland, in his official history, was obviously uncomfortable with the linear model. His discomfort is reflected in his accounts of several efforts by NACA, the air services, and private industry to partition their respective areas of responsibility. He noted, for example, that the "distinction the air forces made between development and evaluation was no more distinct than the NACA line between research and development" (Roland 1985, vol. 1, p. 216). Also, "the term fundamental research was a study in compromise, more suited to blurring boundaries between research functions than clarifying them" (Roland 1985, vol. 1, p. 234).

Sources: In preparing this box I have drawn heavily on the work of William H. Cook (1991), Robert Von der Linden (1991), and John D. Anderson (2002).

[a] Public Law 271, 63rd Congress, 3rd session, passed 3 March 1915 (38 Stat. 930). Reproduced in Roland (1985, vol. 2).

[b] Prandtl regarded his research as applied physics. At Caltech von Karman combined his scientific training with a strong engineering orientation. He arranged for the construction of a wind tunnel and conducted wind tunnel experiments. Beginning in 1932 he became involved in research for Douglas Aircraft, located in nearby Santa Monica, in the developments that led to the DC-3 (Rosenberg 2001).

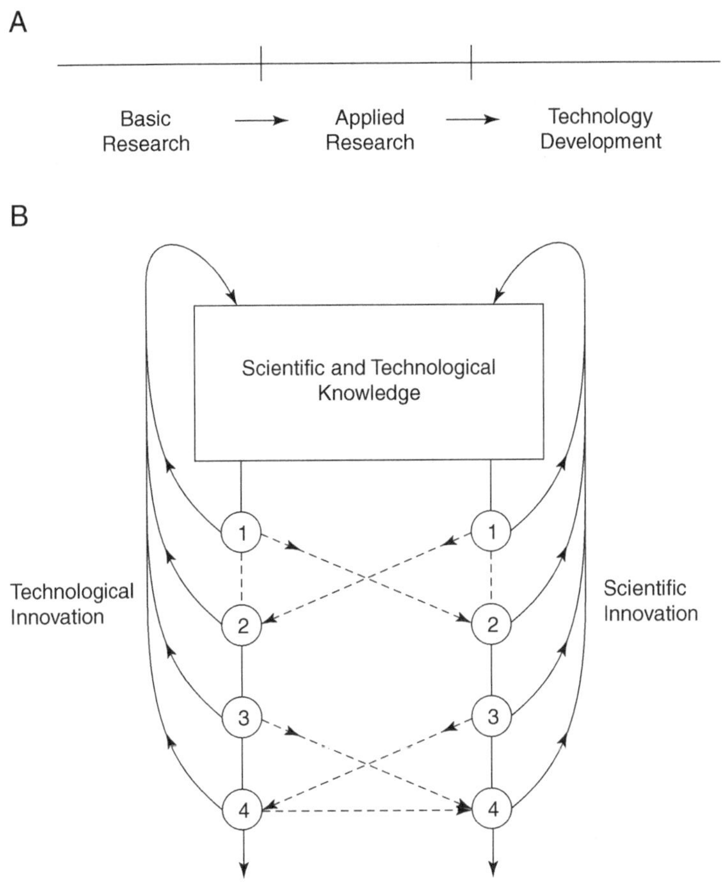

Figure 3.1. Models of the relations between scientific and technical knowledge, and development. (A) Linear model. (B) Interactive model. (C) Quadrant model. *Source:* Vernon W. Ruttan, 2001, *Technology, Growth, and Development: An Induced Innovation Perspective* (New York: Oxford University Press), 81, 537.

War II to address the fundamental problems of supersonic flight, in an attempt to redress its wartime neglect of the potential military and commercial significance of jet aircraft. An important element in this reorientation was a proposal to establish a new Supersonic Research Center dedicated to the development and field testing of supersonic aircraft, missiles, and pilotless aircraft. A central component of the program would be a series of wind tunnels designed specifically to investigate high-speed and high-altitude flight problems, such as structural design, compress-

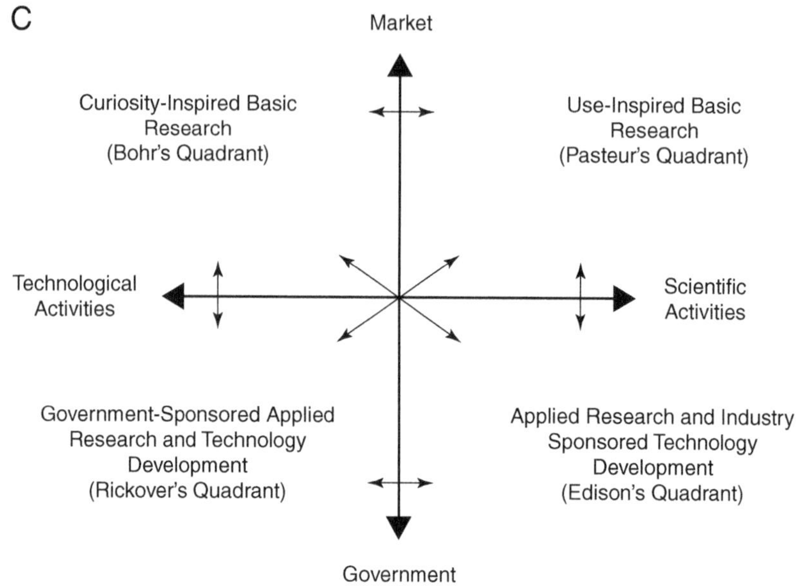

Figure 3.1. (*continued*)

ibility, and stabilization control (Bright 1978, p. 114; Roland 1985, vol. 1, pp. 211–221).[20]

When NACA entered the postwar era, it was also confronted by several difficult institutional problems. One was the determination of the aircraft industry, then America's largest industry, to obtain greater representation on NACA's main governing committees and on the subcommittees that had oversight over NACA research programs. A second was the determination of the Air Force to never again be dependent on NACA as a primary source of aeronautical research and development. A third was the emergence of Congressman Albert Thomas (R-TX) as chair of the Independent Offices Subcommittee of the House Appropriations Committee. Thomas insisted that the committee form of organization of NACA's governance was inappropriate for a government organization of its size.[21] He also ob-

20. The progress of the 1946 designation of NACA as the agency responsible for coordinating all U.S. supersonic research to passage of the Unitary Wind Tunnel Act of 1949 has been characterized as "a tale of Byzantine intricacy" (Roland 1985, vol. 1, p. 217). The budget authorized in the 1949 act was reduced from the original proposal of $500 million to $136 million. From its grand mandate to be the agency coordinating all supersonic research in the United States, "the NACA had been reduced to 'house keeping' for the commercial aircraft industry" (Roland 1985, vol. 1, p. 219).

21. NACA's independent committee form of governance was criticized in 1949 by the Hoover Commission on Organization of the Executive Branch of the Government. The commission recommended that NACA be incorporated into the U.S. Department of Commerce. The Bureau

jected to NACA's size. During his first term he succeeded in reducing the wind tunnel authorization of $136 million to an appropriation of $75 million.

In 1947 Hugh Dryden, then associate director of the National Bureau of Standards, was selected to fill the position of NACA Director of Aeronautical Research. Among the most pressing issues that Dryden had to confront was the direction of NACA research in the post–World War II period. High-speed flight, as noted, was the area in which NACA had the clearest mandate. A high-speed flight program would involve research on instrumentation and on fundamental design concepts. It was not immediately apparent, however, whether the NACA staff that had been assembled to respond to the more immediate flight engineering problems on which NACA had focused during the war had either the training or skills to address more advanced aeronautical research problems that confronted the aircraft industry.

The initial project of the high-speed flight program was the design and development of supersonic aircraft. The project involved close cooperation among NACA, the military services, and the aircraft industry. The NACA-military-industry collaboration was so successful that the "sound barrier" was broken by the Bell X-1 research aircraft on October 14, 1947. The breaking of the sound barrier was the culmination of over twenty years of research at NACA (Hallion 1997; Anderson 2002, pp. 312–314).[22] It brought NACA its third Collier Trophy award, popular and political support, and worldwide acclaim from the aeronautical community. The project also contributed to the acquisition by NACA of the resources it needed to address fundamental problems of stability and control that confronted successful supersonic flight.[23]

In the early 1950s the NACA staff was recognized for a number of fundamental contributions. Richard Whitcomb received the 1954 Collier award for discovery and verification of the area rule of aerodynamic flow that established the optimal relationship between the girth of the fuselage and the wing surface in transonic

of Budget had earlier criticized the committee form of organization on the same grounds (Roland 1985, vol. 1, pp. 221–223). The design of organization charts with direct links to the executive was a favorite preoccupation of students of public administration in the 1940s and 1950s.

22. The breaking of the sound barrier was a public-relations, as well as technical, success. Before the breaking of the sound barrier, it was commonly believed, though not by most aeronautical engineers, that there was a physical barrier to flight faster than the speed of sound. Clarence Johnson, design chief at Lockheed Aircraft Corporation, characterized it as "an engineering stunt costing millions of dollars that could have been more profitably spent on other research projects" (Roland 1995, p. 250).

23. To conduct the flight tests, NACA located a small liaison team at Muroc Air Base in southern California. The group became the NACA High Speed Flight Research Station in 1949 and the High Speed Flight Station in 1954. It also resulted in the construction of a new generation of wind tunnels designed to address the problems of high-speed flight (Roland 1985, vol. 1, pp. 251–252).

flight. This fundamental discovery had an immediate impact on design. Calculations based on the discovery were employed in making the design adjustments needed to enable the first Air Force operational supersonic fighter aircraft to break the sound barrier (Roland 1985, vol. 1, p. 281).

In spite of these successes, NACA continued to allocate a relatively high percentage of its research resources to service-related investigations. This was not because NACA's influential military service and industry clients were oblivious to the need for fundamental research, but because of pressures to get the next generation of fighter aircraft into operation or the prototype of the next commercial airliner into production (Roland 1985, vol. 1, p. 288).

By the mid-1950s NACA "had gone a long way toward restoring its reputation and dimming the memories of how it had been bested by the Germans during the War" (Roland 1985, vol. 1, p. 255). After three years of budget reverses, President Dwight D. Eisenhower approved a budget supplement in 1955 and regular increases for the next three years. When the Soviet Union (Union of Soviet Socialist Republics) launched Sputnik I on October 4, 1957, it set in motion a series of events that would lead to NACA's being absorbed into a new agency, NASA (see chapter 7).

Military and Commercial Aircraft at Boeing

The relationship between military procurement and commercial aircraft development is illustrated with particular force in the history of Boeing Aircraft and Transport Company (now Boeing Aircraft Company). During the 1920s and 1930s Boeing was responsible for a number of advanced technical innovations, but it experienced difficulty in achieving consistent success in the commercial market. Only reliance on postal and military contracts enabled Boeing to remain economically viable in the unstable markets for military and commercial aircraft in the interwar period. During the 1940s and into the 1980s, Boeing was able to translate its successful development of a series of military aircraft into a dominant position in the United States and world commercial aircraft market. This dominance began to erode in the late 1980s.

Wartime Success

In 1931 Boeing, drawing on its own resources, initiated a program to develop a revolutionary new bomber, the B-9, which incorporated advances in aircraft design that had accumulated for over a decade.[24] Boeing also initiated a program to develop a new commercial airliner, the B-247, based on the B-9. The B-247 was the first high-performance multiengine commercial airliner. It established the configuration of the commercial airliner until the introduction of jet airliners in the late 1950s. It was introduced, however, before a number of technical advances, such as the NACA Cowling and the variable-pitch propeller, became available. Internal corporate disagreement resulted in a plane with relatively small passenger capacity.[25]

Boeing bombers played an important role in World War II. By the end of World War II, 13,726 B-17 bombers had been built. When in January 1940 the Air Force put out another request for proposals for an even higher performance bomber, Boeing already had studies under way looking toward a successor to the B-17. The B-29 was the first Boeing plane in which significant design improvements were made in accordance with wind tunnel data obtained during construction. It was the only wartime airplane to have cabin pressurization. "During World War II the B-29 ferried supplies from India over the Himalayas to China, it flew bombing missions from the Marianas to Japan, and it carried the atomic bombs that were dropped on Hiroshima and Nagasaki" (Cook 1991, p. 80).

Boeing was only partially successful in its attempts to develop economically successful commercial aircraft based on its successful wartime bombers. Production of the Boeing 307 Stratoliner, the first commercial airliner with a pressurized cabin, derived from the B–17. Production was discontinued in 1939, shortly after it was introduced, because of the pressure of military activity. Only ten were built. After the war Boeing built the Stratocruiser, a long-range commercial transport derivative of the B-50 postwar bomber and the C-97 refueling tanker. It was powered by a Pratt and Whitney 4360 engine—one of the largest piston engine motors ever installed in a production airliner. Very few were ever sold. Cook suggests that problems with the engine prompted early adoption of the jet engine by Boeing (1991, p. 58). Production of the Stratocruiser was discontinued in 1950.

24. In this section I draw heavily on Von der Linden (1991).

25. Although Boeing set the stage—the B-247 has been characterized as the first modern airliner—it was Douglas, as noted earlier, that succeeded in translating the concept into the world's most successful commercial airliner. Douglas moved very rapidly to complete and test a DC-1 prototype. It made its first flight on July 1, 1933—only five months after Boeing delivered its first B-247 to United. "The DC-1 prototype performed so well," comments Cook, "that the production version, the DC-2 was ordered by all U.S. airlines except United" (1991, p. 54). The first delivery of the DC-2 occurred in May 1934, and that of the first DC-3 in December 1935 (Cook 1991, pp. 50–51): "Thirteen thousand DC-3's and the military equivalent, the C–47, were built, and were eventually used all over the world. . . . They were operated indefinitely and wore out a multitude of engines" (Cook 1991, p. 55).

The Design Revolution

The end of World War II in 1945 left Boeing, and the aircraft industry generally, with considerable uncertainty about the future.[26] At the end of the war, Boeing had under way development work on a straight-wing two-engine bomber with engine nacelles tucked under the wing. Models were already being wind tunnel tested by NACA at Langley Field when in 1945 a survey team of U.S. scientists and engineers discovered at the German aircraft research facility at Braunschweig, Germany, an exhaustive body of aerodynamic data on tests of swept-wing aircraft. The team included George Schairer, a young aeronautical engineer from Boeing, who brought the swept-wing designs to the attention of the team working on the development of the XB-47.

After evaluating the German data, Boeing proposed to the Air Force that the design be changed to swept-wing configuration. Air Force engineers at Wright Field insisted that two engine pods be mounted under each wing. In April 1946, when agreement had been reached on design, the Air Force project office at Wright Field was sufficiently impressed to sign a contract for two XB-47 prototypes at $10 million each (Cook 1991, pp. 148–170; Anderson 2002, pp. 290–292). In September 1947 the first XB-47 prototype was rolled out of its hangar. The first flight occurred on December 17, 1947.

Following completion of flight tests by Boeing and the Air Force, General K. B. Wolf, head of bomber production at Wright Field, was persuaded during a visit to Seattle to stop by Moses Lake to look at the XB-47 and to take a short ride on the plane. General Wolf had never flown in a jet before:

> The flight, piloted by Major Townsend, was comprised of a climb up to altitude for a level speed run then a low pass by the tower followed by a steep climb. . . . After the flight General Wolf was visibly elated and asked Major Townsend if he considered the airplane to be operational. When Townsend answered "Yes" General Wolf almost immediately responded that the Air Force would buy it and that it would be built "as is. . . ." General Wolf thus revolutionized strategic bombing after only a 20 minute ride. (Cook 1991, p. 194)

The B-17 became the U.S. Strategic Air Command's first viable and successful jet bomber.

26. In this section I draw heavily on Cook (1991). See also Anderson (2002, pp. 283–359). Between 1945 and 1948 military procurement dropped dramatically. Each of the fifteen active aircraft producers ran substantial deficits. Aircraft manufacturers attempted to diversify into products such as buses, kitchen ranges, prefabricated homes, home freezers, and automobile and marine engines (Crouch 2003, pp. 488–494). I myself still own two practically indestructible Grumman canoes.

As wind tunnel testing of the XB-47 progressed during 1945–1949, a number of Boeing engineers began to consider the prospect of developing a commercial jet transport.[27] It was considered doubtful, however, that initial sales could justify development costs. It did become apparent, however, that a military jet tanker specifically designed to refuel the B-52 could represent not only an economically attractive opportunity for Boeing, but also an initial step toward the development of a commercial jet airliner. Steps were initiated to develop a prototype tanker—the Dash-80.[28]

The Dash-80 was rolled out of the hangar in May 1954 for ground testing. Several defects, primarily in the landing gear, delayed flight testing until mid-July. Initial flight testing revealed several more defects, which were quickly corrected. "The early flights of the Dash-80 demonstrated good flying characteristics," as Cook concludes. "The overall configuration was judged highly successful" (1991, p. 221). The Air Force's decision to purchase the KC-135 tanker, the production version of the Dash-80, resolved the question of financing a commercial jet transport version. On September 8, 1952, Boeing made a public announcement that it had initiated the development of a commercial jet transport—the Boeing 707. Boeing decided to "bet the company" on the future of commercial jet aircraft.

On September 8, 1952, five months after the detailed design work had been started, Boeing made a public announcement that work on a jet transport was under way. Most firms in the airline industry remained skeptical of the potential technical and economic advantages of a commercial jet airliner. Although it would fly higher and faster than the piston-propeller transports, it was less fuel efficient, and engine maintenance problems remained a major concern.[29]

On October 13 Pan American ordered twenty Boeing 707s. It also ordered twenty-five DC-8s, Douglas Aircraft's announced entry into the commercial jet

27. In this and the next several paragraphs, I draw primarily on Cook (1991, pp. 211–226).

28. It was estimated that the cost of constructing the prototype would be in the $16 million range. However, it would be possible to charge a considerable part of the cost to the government under the Independent Research and Development provisions of Boeing's Air Force contracts. As a result, the direct cost to Boeing was estimated at only $3 million (Cook 1991, p. 214).

29. U.S. engine manufacturers were also slow to realize the market potential for turbojet engines. Even while Boeing engineers were pressing Pratt and Whitney for advanced turbojet engines for the KC-135 military tanker, Pratt and Whitney management still considered the turboprop as the engine of the future. In the mid-1950s it took pressure from American Airlines management on Boeing, and Boeing pressure on Pratt and Whitney, to develop the fan-jet engines American wanted installed in a new version of the Boeing 707. Cook comments that in 1953 Pratt and Whitney had little feel for the changes in design studies going on in the airline industry or their implications for engine manufacture (1991, p. 229). For a definitive history of aircraft gas turbine engine development in the United States, see St. Peter (1999).

market.[30] On November 8 American Airlines placed an order for thirty 707s on the condition that Boeing would produce a wider version of the aircraft. "The American order for 707s was absolutely central to Boeing's future in the commercial jet transport industry," explains Cook, "because it provided an endorsement that influenced many other airlines to follow American's lead" (1991, p. 243). The greater width that American insisted on was important in establishing the Boeing policy of developing a "family" of closely related passenger aircraft for different services (table 3.2).

The Boeing 707 set the standard for modern commercial aircraft design. It also confirmed the success of the Boeing policy of using military aircraft contracts to test its new concepts and to provide funding for development work that would feed into the design of its commercial aircraft. It established Boeing as the world's leading producer of commercial aircraft.

However, the development of the Boeing 747 followed a somewhat different pattern. In 1965 Boeing lost to Lockheed an Air Force competition to design a large military transport. The Lockheed plane, the C-5A, was considered both too large and too slow to develop into a commercial transport. Boeing engineers, starting with the design they had developed for the military transport, proposed the development of an even larger commercial jet. The plane that emerged from this effort, the Boeing 747, set the design standard for all subsequent wide-bodied jets (Anderson 2002, p. 354). The 747 was not, however, an immediate commercial success. The American economy was in the midst of a recession. Although it had received an initial order from Pan American in 1966 and orders from KLM and Lufthansa in 1971 and 1972, there was a period of eighteen months in the early 1970s when Boeing did not receive a single order from a U.S. domestic airline (Crouch 2003, p. 628). The Boeing 767, launched in 1987, and the Boeing 777, launched in 1998, were evolutionary from the 747 (table 3.2).

By the mid-1990s it seemed clear that military contracts would no longer play a significant role in the development of U.S. commercial airliners. The defense aircraft industry had become increasingly specialized. The military's unique performance requirements, such as stealth and maneuverability, have forced separation of the military from the commercial suppliers: "Boeing, America's major commercial supplier, is not the prime contractor for any Department of Defense aircraft (either fighters or bombers) while Lockheed, which now does no commercial work,

30. Juan Trippe, the founder and still president of Pan American, had consistently insisted on maintaining a preeminent position within the airline industry by being the first to put new airliners into service. Pan American was the first to buy the Douglas DC-6, the Lockheed Constellation, and the Boeing Stratocruiser (Cook 1991, p. 236).

Table 3.2. The Boeing Family of Commercial Passenger Aircraft from the 707 to the 777

Name	Date Introduced	Remarks
707	1957	First American commercial jet transport. Established configuration for the next generation of commercial jets.
727	1964	First tri-jet introduced to commercial service. Shorter, with slower takeoff and landing speed than 707. Could be accommodated by smaller airports.
737	1957	A short- to medium-range plane. Wing extensions (winglets) contributed to an increase in fuel efficiency and range.
747	1969	First wide-bodied commercial jet transport. Established the design standard for all subsequent wide--bodied jets. Boeing 747-400 is the cargo freighter with the lowest cost per ton mile.
757	1982	A medium-range narrow body successor to 727 with the same cabin width but a longer fuselage.
767	1981	Narrowest wide-bodied jet. Developed in tandem with 757.
777	1998	Larger than other twin or tri-jets, and smaller than 747. First jetliner to be 100 percent digitally designed. Airplane was "preassembled" on the computer.

Source: J. D. Anderson, Jr., 2002, *The Airplane: A History of Its Technology* (Reston, VA: American Institute of Aeronautics and Astronautics). Reprinted with permission of the American Institute of Aeronautics and Astronautics. Boeing Aircraft, http://boeing.com/commercial/707family, as well as for other members of the Boeing family of commercial passenger aircraft.

is the prime contractor for the next generation of military fighter aircraft" (Gansler 1995, p. 35).

The NASA Era

The launching of Sputnik I and Sputnik II by the Soviet Union generated substantial concern in the media, among the general public, and among members of Congress that the United States was lagging behind the Soviet Union in the "space race" (see chapter 7). One response was to absorb NACA, its laboratories, and its employees into a newly established NASA. NASA was authorized "to carry out the design, development, and testing of aeronautical and space vehicles (Bromberg 1999, p. 17). In addition to taking over the facilities and functions of NACA, NASA absorbed over the next several years the Jet Propulsion Laboratory, the Army Ordinance Ballistic Missile Agency, the civilian space program previously managed by the Department of Defense Advance Research Project Agency (ARPA), and the other space programs of the military services (figure 3.2). Although NASA was established to provide civilian leadership of the space program, it retained very substantial defense-related functions (see chapter 7).

Aeronautics Research at NASA

The incorporation of NACA into NASA resulted in a substantial shift from research conducted primarily in-house—the armory system—to research conducted by politically powerful aerospace contractors (Bromberg 1999, pp. 15–44). It also resulted in a massive redirection of NACA/NASA R&D from aeronautic to space-oriented R&D.[31] The major burden of research and technology development to meet continuing military needs was transferred to the U.S. Department of Defense (DOD)—although at a significantly reduced level (OSTP 1982, p. 14).[32]

The U.S. Army Air Force emerged from World War II committed to a doctrine of the efficiency of strategic bombing and development of bombers with global range. The chosen instrument of the military was at first the Boeing B-36, with

31. The late 1950s marked the transition of the aircraft industry to the aerospace industry. The term *aerospace* includes development and production of manned and unmanned aircraft, missiles, propulsion systems, space vehicles, and associated electronics (Pattillo 1998). I discuss the emergence of the space industries in chapter 7.

32. In some NASA literature the abbreviation *R&T* is used to refer to *research and technology development*. I regard *R&T* a more accurate abbreviation than *R&D*. However, for consistency, I continue to use the *R&D* appellation.

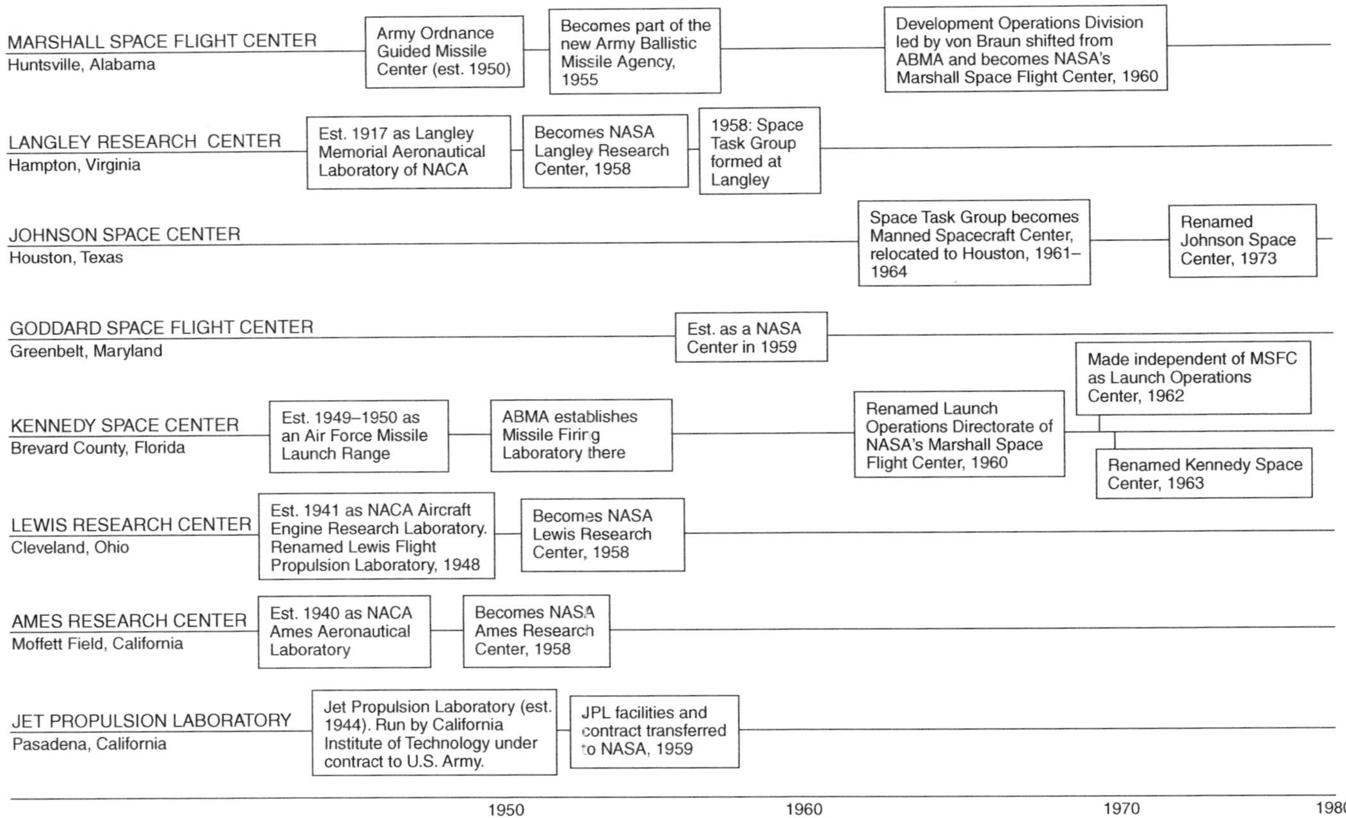

Figure 3.2. Major NASA installations. The NASA centers were made up of NACA installations, laboratories transferred from the Armed Services, and a facility, the Goddard Space Flight Center, built specifically for the new agency. *Source:* Joan Lisa Bromberg, 1999, *NASA and the Space Industry* (Baltimore: Johns Hopkins University Press), 10–11. Copyright © 1999. Reprinted with permission of The Johns Hopkins University Press.

its six pistons and four jet engines, introduced in 1949. By the time the B-36 was introduced, the Air Force was already committed to full conversion to jet engines. This emphasis was by then fully consistent with the objectives of the commercial airline industry. As already noted, the Boeing airframe continued to evolve to the B-47 and to the B-52, and from the 707 to the 747 families of commercial aircraft. Boeing set the pattern for the development of long-distance commercial aircraft throughout the world from the late 1950s to the early 1990s (see table 3.2).

Defense-related R&D, including spending by the Air Force, in addition to that performed by NACA/NASA and the funding of research conducted by the aircraft industry itself, remained remarkably stable. In 1960 R&D spending in the aerospace industry, including the aircraft industry, accounted for about 20 percent of gross revenues and remained in this range through the 1960s and 1970s. Of this amount only 10.3 percent, or approximately 2 percent of gross revenue, was privately financed. The percentage privately funded had by 1978 risen to approximately 5 percent of gross revenue—high by historical standards in the aircraft industry, but low in comparison with other high-technology industries (Bluestone, Jordon, and Sullivan 1981, pp. 158–159). The military and space programs continued to account for at least two thirds of R&D directed to increasing aircraft performance until at least the early 1980s (Mowery and Rosenberg 1982b, p. 171; Mowery and Rosenberg 1989).

By the early and mid-1980s, it was becoming clear that technology transfer from military procurement was no longer a dynamic source of technical change in the commercial airline industry. There has been very little direct spin-off from military to commercial aircraft since the development of stealth aircraft beginning in the mid-1970s. The effect of deregulation of the aircraft industry had the effect of weakening commercial demand for new aircraft that could not offer a promise of dramatic operating cost reductions. Both the military and commercial aircraft industries were beginning to show signs of technological maturity (see chapter 8).

International Competition

From the early 1980s, government concern with the U.S. competitive position in the international commercial aircraft industry, stimulated by European efforts to strengthen their own commercial aircraft industry, led to a substantial increase in NASA expenditures directed to commercial aircraft technology development. By the early 1990s NASA and the Air Force were engaged in R&D in almost every dimension of aircraft technology and were devoting resources, variously estimated to be upward of $1 billion per year, to large commercial aircraft research, development, and demonstration (Pattillo 1998, p. 344; Lawrence 2001). Although the aircraft industry was regarded by many as a relatively mature industry, productivity growth, measured by partial indicators such as operating cost per seat mile, continued to improve (Miller and Sawers 1968; Mowery and Rosenberg 1982a).[33]

A policy issue that has confronted public support for aeronautical R&D since NACA was established has been how to distinguish the R&D that is properly supported by public resources from that supported by private. There are clearly structural characteristics in technology development in the aircraft industry that support an argument for a substantial public-sector role. Aeronautical R&D is characterized by very capital-intensive research facilities—wind tunnel, flight test, propulsion, and other special capability facilities. In the past the close articulation between R&D for military and that for commercial purposes assured that a definitive answer to this issue was seldom attempted and has proven elusive when addressed.

Consolidation of the civil large-bodied aircraft industry, in which Boeing and Airbus have become the only significant players, and of the military aircraft industry, in which Lockheed-Martin has become the most significant U.S. player, suggests that this is an appropriate time to revisit the implications of the changing relationship between public support for defense-related R&D and commercial aircraft R&D. For more than a decade, the direct and indirect subsidies to commercial aircraft development in Europe and the United States have been the subject of intense international trade negotiations and of a large but inconclusive polemical literature (Fisher 2002; "Toward the Wild Blue Yonder," 2002).[34]

33. Until well into the 1970s, both military and commercial aircraft performance continued to be characterized primarily in terms of altitude and speed. The culmination of this trajectory in commercial aircraft was the French-British decision to build the Concorde, and the U.S. decision to build an American supersonic transport (SST). Only twenty Concordes were produced. The plane has been characterized as a technical success but a commercial failure. The last Concorde regular transatlantic flights were concluded in 1993. After much political acrimony, development of the U.S. SST was aborted. Since the early 1970s developments of commercial and military aircraft have followed different trajectories. The focus of commercial aircraft development has shifted to operational efficiency. Since the mid-1990s two producers of small, fuel-efficient jetliners, designed to capture markets linking smaller cities with major airline hubs, have been introduced by Bombardier, of Canada, and Embraer, in Brazil (Christensen, Anthony, and Roth 2004, pp. 130–150).

34. In 1990 Airbus Industries exceeded Boeing in the sales of every class of airplane smaller than the 747. In 2003 Airbus deliveries of wide-bodied jets exceeded deliveries of Boeing wide-bodied jets for the first time. Airbus announced that it was designing a "superjumbo" jet for delivery in 2006, while Boeing announced plans for a more fuel-efficient successor to the Boeing 767. Meanwhile, the DOD was engaged in an internal battle about a convoluted arrangement under which Boeing would build 100 large tanker aircraft, which the Air Force would lease from Boeing (Jehl 2003).

Perspective

The private sector played a primary role in the initial development of aircraft. The Wright brothers self-financed the invention and initial development of the first successful propeller-driven aircraft. During the next decade design and technical improvements were made primarily by flight enthusiasts, craftsmen, and engineers. World War I played a major role in inducing the development of an aircraft industry in both Europe and the United States, but it was not, with the exception of the all-metal airplane, productive of revolutionary developments in aircraft technology and design. The airplanes that flew during World War I were evolutionary from the Wright Flyer.

During the interwar period, advances in engineering and aeronautical knowledge established a more solid foundation for aircraft design in both Europe and the United States. The initial research and development of jet aircraft engines in the 1930s by Whittle in Britain and von Ohain in Germany were supported from their own resources. By the late 1930s, however, public support for R&D, largely motivated by military demand, played a dominant role in jet aircraft technology development.

It is hard to avoid the conclusion that, in the interwar period, commercial aircraft would have been developed and introduced more slowly in the absence of defense-related technology development and military procurement. In the United States, NACA played a particularly important role in the development of dual-use technology applicable to both military and commercial aircraft. It is an open question whether, in the absence of NACA research and military procurement, propeller-driven commercial aircraft would have achieved the level of technological maturity represented by the DC-3 by the mid-1930s (Mowery 2004). It also seems apparent that delay in the development of a mature piston-driven technology would have reduced the urgency of the development and introduction of jet aircraft. There can be no question, however, that the advances in aircraft design represented by the Boeing 707 and 747 would have been substantially delayed in the absence of the stimulus provided by military procurement. The strategy employed by Boeing in using military aircraft contracts to test new concepts and to fund the development work that fed into the design of advanced commercial aircraft would not have been available.

What about the future? Shortly after the end of World War II, it became apparent to knowledgeable scientists, engineers, and industry leaders that a transition would need to be made from the piston-propeller to a jet-propulsion aircraft trajectory of aircraft development. My reading of the literature today suggests that, in spite of optimistic rhetoric about the future of aviation, there has not yet emerged even the beginning of a consensus about the parameters of a next revolution in either military or commercial aircraft design.

A question that can hardly be avoided, as one reflects on the history of public support for research and technology development and the support provided by military procurement for the U.S. aircraft industry, is whether a fully private U.S. commercial aircraft industry can ever be economically viable. Two decades ago Barry Bluestone, Peter Jordan, and Mark Sullivan raised the same question and answered it negatively: "Without federal government [support] there would simply be no aircraft industry despite the fact that the commercial market is playing a much larger role than it has in the past. No aspect of the industry, including the commercial sector, could exist without the R&D funds provided by the state, or the state's purchase of military equipment" (1981, p. 170).

The wide-bodied commercial aircraft industry continues to introduce new models at prices well below marginal costs, in the expectation that production will become profitable as they work their way down their learning curves. The price of the Lockheed L1011, designed to compete with the Boeing 747 and the Douglas DC-10, remained below its marginal cost for its entire fourteen-year production run (Benkard 2004). The 2001 commitment by Airbus to the development the Airbus A-380 (A-3xx) superjumbo and the subsequent decision by Boeing not to proceed with the development of the Sonic Cruiser, a transonic passenger transport, are not inconsistent with the Bluestone hypothesis. In late 2003 Boeing announced that it would begin seeking orders for a smaller, privately financed 7E7 "Dreamliner" plane, which it would begin to deliver sometime in 2008 (Linsford 2003).

Regardless of the precision of my counterfactual speculations, it is clear that the world of aeronautics and aviation would be far different today in the absence of military procurement—far different than if aircraft development had depended entirely on private-sector investment in research and technology development, and on commercial aircraft demand. In the United States, military considerations were largely responsible for the establishment and funding of aeronautics research at both NACA and NASA. The development of jet propulsion technology in Germany and Britain, as well as its transfer to the United States, was largely a product of military demand. Without the support of technology development and military procurement, it is doubtful that the advances in design and technology represented by the wide-bodied jet aircraft would yet dominate long-distance military or commercial transport.

ACKNOWLEDGMENTS

I am indebted to David Mowery, Alex Roland, Nathan Rosenberg, and Walter Vincenti for comments on an earlier version of this chapter.

REFERENCES

Anderson, J. D., Jr. 2002. *The Airplane: A History of Its Technology*. Reston, VA: American Institute of Aeronautics and Astronautics.

Benkard, C. L. 2004. "A Dynamic Analysis of the Market for Wide-Bodied Commercial Aircraft." *Review of Economic Studies* 71: 581–611.

Bluestone, B., P. Jordan, and M. Sullivan. 1981. *Aircraft Industry Dynamics: An Analysis of Competition, Capital, and Labor*. Boston: Auburn House.

Bobo, E. 2001. "Scientists at War: The Development of Radar and Jet Propulsion in Britain." In *War in the Age of Technology: Myriad Faces of Modern Armed Conflict*, ed. G. Jensen and A. West. New York: New York University Press, 239–262.

Bonaccorsi, A., and P. Giuri. 2000. "When Shakeout Doesn't Occur: The Evolution of the Turboprop Engine Industry." *Research Policy* 29: 847–870.

Bright, C. D. 1978. *The Jet Makers: The Aerospace Industry from 1945 to 1972*. Lawrence: Regents Press of Kansas.

Bromberg, J. L. 1999. *NASA and the Space Industry*. Baltimore: Johns Hopkins University Press.

Christensen, C. M., S. D. Anthony, and E. A. Roth. 2004. *Seeing What's Next? Using the Theories of Innovation to Predict Industry Change*. Boston: Harvard Business School Press.

Constant, E. W. 1980. *Origins of the Turbojet Revolution*. Baltimore: Johns Hopkins University Press.

Cook, W. H. 1991. *The Road to the 707: The Inside Story of Designing the 707*. Belleview, WA: TYC Publishing.

Crouch, T. D. 2003. *Wings: A History of Aviation from Kites to the Space Age*. New York: Norton.

Fisher, D. I. 2002. "Super-Jumbo Problem: Boeing, Airbus, and the Battle for the Geopolitical Future." *Vanderbilt Journal of Transnational Law* 35: 865–893.

Gansler, J. S. 1995. *Defense Conversion: Transforming the Arsenal of Democracy*. Cambridge, MA: MIT Press.

Gorn, M. H. 2001. *Expanding the Envelope: Flight Research at NACA and NASA*. Lexington: University Press of Kentucky.

Gray, G. W. 1948. *Frontiers of Flight: The Story of NACA Research*. New York: Knopf.

Hallion, R. P. 1977. *Legacy of Flight: The Guggenheim Contribution to American Aviation*. Seattle: University of Washington Press.

———. 1997. *Supersonic Flight: Breaking the Sound Barrier and Beyond: The Story of the Bell X-1 and Douglas D-558*. London: Brassey's.

Hanley, P. 1982. *Bringing Aeronautics to America*. Cambridge, MA: MIT Press.

Heppenheimer, T. A. 1995. *Turbulent Skies: The History of Commercial Aviation*. New York: Wiley.

Heron, S. D. 1961. *History of the Aircraft Piston Engine: A Brief Outline*. Detroit, MI: Ethyl.

Higham, R. 2003. *One Hundred Years of Air Power and Aviation*. College Station: Texas A&M University Press.

Holley, I. B., Jr. 1964. *Buying Aircraft: Materiel Procurement of the Army Air Forces*.

Washington, DC: U.S. Government Printing Office for Department of the Army, Office of the Chief of Military History.

Hunsaker, J. C. 1952. *Aeronautics at the Mid-Century*. New Haven, CT: Yale University Press.

Jehl, D. 2003. "Air Force Pursued Boeing Deal Despite Concerns of Rumsfeld." *New York Times*, 6 December, p. A1.

Lawrence, P. K. 2001. *Aerospace Strategic Trade: How the U.S. Subsidizes the Large Commercial Aircraft Industry*. Aldershot, U.K.: Ashgate.

Linsford, J. L. 2003. "Boeing to Hunt for 7E7 Orders: A Step toward Plane's Launch." *Wall Street Journal*, 17 December, B-21.

Miller, R., and D. Sawers. 1968. *The Technical Development of Modern Aviation*. London: Routledge and Kegan Paul.

Mowery, D. C. 2004. Letter to Vernon Ruttan. August 11.

Mowery, D. C., and N. Rosenberg. 1982a. "The Commercial Aircraft Industry." In *Government and Technical Progress: A Cross Industry Analysis*, ed. R. Nelson. New York: Pergamon Press, 191–161.

———. 1982b. "Technical Change in the Commercial Aircraft Industry." In *Inside the Black Box: Technology and Economics, 1925–1975*, ed. N. Rosenberg. Cambridge: Cambridge University Press.

———. 1989. *Technology and the Pursuit of Economic Growth*. New York: Cambridge University Press.

Office of Science and Technology Policy (OSTP). 1982. *Aeronautical Research and Technology Policy*, Vol. 1. Washington, DC: Executive Office of the President.

Pattillo, D. M. 1998. *Pushing the Envelope: The American Aircraft Industry*. Ann Arbor: University of Michigan Press.

Roland, A. 1985. *Model Research: The National Advisory Committee for Aeronautics: 1915–1958*, vols. 1 and 2. Washington, DC: National Aeronautics and Space Administration.

Rosenberg, N. 2001. "Engineering Knowledge." Palo Alto, CA: Stanford University Department of Economics. Mimeograph.

Ruttan, V. W. 2001. *Technology, Growth, and Development: An Induced Innovation Perspective*. New York: Oxford University Press.

Sahal, D. 1981. *Patterns of Technological Innovation*. Reading, MA: Addison-Wesley.

Simonson, G. R. 1960. "The Demand for Aircraft and the Aircraft Industry." *Journal of Economic History* 20: 361–382.

St. Peter, J. 1999. *The History of Aircraft Gas Turbine Engine Development in the United States: A Tradition of Excellence*. Atlanta, GA: International Gas Turbine Institute.

"Toward the Wild Blue Yonder: Boeing vs. Airbus." 2002. *Economist*, 27 April, pp. 67–70.

Trimble, W. F. 1986. "The Naval Aircraft Factory, the American Aircraft Industry and Government Competition, 1919–1928." *Business History Review* 60: 176–198.

U.S. Bureau of the Census. 1960. *Historical Statistics of the United States: Colonial Times to 1957*. Washington, DC: U.S. Government Printing Office.

Vander Meulen, J. 1992. "Political Culture, Public Policy, and the Development of the American Aircraft Industry to 1945." *Business and Economic History* 21: 56–64.

Vincenti, W. G. 1990. *What Engineers Know and How They Know It: Analytical Studies from Aeronautical History*. Baltimore: Johns Hopkins University Press.

Von der Linden, F. R. 1991. *The Boeing 247: The First Modern Airliner*. Seattle: University of Washington Press (for National Air and Space Museum).

———. 2002. "Airlines and Air Mail: The Post Office and the Birth of the Commercial Aviation Industry." Lexington: University Press of Kentucky.

4

Nuclear Energy and Electric Power

Before the nineteenth century the primary sources of energy were animal and human power, fuel wood and agricultural wastes, and wind power and water power. The industrial revolution involved two major transitions in energy use. The first was made possible by the steam engine, the first technology for the conversion of fossil energy resources into work. The second was the introduction of electricity, the first energy carrier that could be converted into light, heat, and work at the point at which it is used (Grübler 1998, pp. 149–251).[1]

These technical changes have been associated with dramatic changes in the relative importance of the several sources of energy. In the nineteenth century, with the expansion of railroads and the growth of the steel and electric power industries, the use of coal for energy rose dramatically. Oil and natural gas were introduced in the 1870s. Their use has been closely associated with the diffusion of the internal combustion engine and the growth of the petrochemical industry. By the 1970s, each accounted for a larger share of energy than coal. Nuclear power use, which experienced exceptionally rapid growth after its introduction in the late 1950s, has experienced slower growth since the early 1980s.

In this chapter my primary objective is to explore the role of the U.S. military and defense-related institutions in the scientific and technical development of nuclear power in the United States. Institutional innovation is an important subtopic. The Manhattan Project, which was organized to produce the atomic bomb, was one of the most significant institutional innovations during World War II. It marked the transition from the public armory system to the private contractor system in

1. In this chapter I draw on a more detailed discussion of the development of the electric light and power industries in Ruttan (2001, pp. 235–285).

the development of armaments and in the transition to "big science" in the mobilization of scientific resources to address mission-oriented research and development (Danhof 1968, pp. 93–99).

The Electric Utility Industry

The initial development of electric power technology took place entirely in the private sector. In 1876 Thomas A. Edison established the first modern industrial laboratory in the United States in Menlo Park, New Jersey. He visualized the laboratory as an "invention factory"—capable of turning out "a minor invention every ten days and a big thing every six months or so." The invention of the high-resistance incandescent lamp and the development of a system for the generation and distribution of electric power by Edison and his research team established the technical foundation for the electric utility industry (Hughes 1983).

Until the advent of nuclear power, the technology of electric power generation would have remained familiar to Edison. Each power generation "unit," called a *boiler-turbogenerator unit,* operated independently of other units. It consisted of a boiler to burn the fuel and to generate and expand the steam, and a turbogenerator to convert high-pressure steam into electric energy. A condenser converted the steam into hot water to complete the cycle. Until the late 1960s technical change was aimed primarily at increasing the size of generators and boilers, and improving the thermal efficiency of the generating cycle (Ruttan 2001, pp. 260–262).

A series of equally important institutional innovations was required to realize the economic gains made possible by the advances in technology and the integration of equipment manufacturers into a coherent electric supply (utility) industry. The manager-entrepreneur Samuel Insull became a pivotal figure in the institutional innovations that gave rise to an integrated publicly regulated electric supply industry in the United States. In 1892 Insull, who had worked with Edison at Menlo Park, moved to Schenectady to become manager of the Edison General Electric Company. When it merged with the Thompson-Huston Company, a leading manufacturer of electrical equipment, to form the General Electric Company, Insull moved to Chicago to become president of the Chicago Edison Company, one of the many small electric power companies that served the Chicago area market. Within two decades Insull and his associates succeeded in creating a single, "monopolistic, technologically efficient, and economically operated company for all Chicago" (Hughes 1989, p. 140).

As the Chicago Edison Company expanded to include most of Illinois, Insull was able to persuade the state legislature to enact legislation that substituted state regulation of rates and service for local regulation.[2] The Illinois system became the model for regulation in other states: "The utility industry interpreted such developments in Illinois and elsewhere as an implicit social contract in which the utilities undertook to provide reliable and affordable electricity in exchange for a socially determined rate of return" (Ruttan 2001, p. 246). Innovations in state and federal regulatory regimes in the 1980s and early 1990s, particularly the Energy Policy Act of 1992, contributed to the deconstruction of the institutional structure that Insull and other leaders of the electrical power industry had erected in the 1920s. These policy reforms enabled independent power producers using natural gas as a primary energy source to walk through the door that energy activists thought they had opened for sustainable energy sources such as biomass, wind, and solar-thermal (Hirsh 1999).

Nuclear Energy

Demonstration of the feasibility of controlled nuclear fission by a team directed by the young Italian physicist Enrico Fermi at the University of Chicago's Stagg Field on December 2, 1942, set the stage for an active role of the U.S. military and defense-related institutions in technology development for the electric power industry.[3] From its beginning, it has not been possible to understand the development of the nuclear power industry apart from the military application of nuclear energy (Cowan 1990; Cantelon, Hewlett, and Williams 1991, pp. 303–307).

Atoms for War

The steps that led to Fermi's demonstration of the possibility of controlled nuclear fission, and a few years later to the construction of the nuclear bomb, were set in

2. In Chicago and elsewhere in Illinois, politics became an essential component of the Insull system: "Methods were found by which politicians obtained wealth from political power without having to steal public money" (Hughes 1983, p. 206).

3. I do not attempt in this section to provide a scientific or technical account of the development of atomic power. My purpose is to provide insight into the critical role played by military considerations in the development of atomic power for military and commercial purposes. Throughout this section I draw heavily on the exceedingly useful official histories by Hewlett and Anderson (1962) and Hewlett and Duncan (1969). See also Laurence (1959), Hughes (1989, pp. 353–44), and Pool (1997). For excellent technical accounts of nuclear fission, nuclear fusion, nuclear reactors, and nuclear weapons, see "Nuclear Fission, Nuclear Fusion, Nuclear Reactors, and Nuclear Weapons" (1974).

motion in 1938 when German chemists Otto Hahn and Fritz Strassman, at the Kaiser Wilhelm Institute in Berlin, found that they could split uranium atoms by bombarding their nuclei with neutrons. The neutron bombardment causes the uranium to "fission" into smaller pieces that fly apart with a great deal of energy.[4]

It was immediately recognized in the physics community, in both Europe and the United States, that if the energy liberated by the splitting of the uranium atom could be controlled and directed, it might be possible to construct a nuclear weapon far more powerful than anything that was currently available (Hewlett and Anderson 1962, pp. 10–14). This possibility led Leo Szilard, a Hungarian physicist who had immigrated to the United States and was then employed at Columbia University, to attempt to bring the implications of the discovery to the attention of the U.S. government. Szilard contacted Albert Einstein about his concern. On August 2, 1939, Einstein signed a letter, prepared by Szilard and addressed to President Franklin D. Roosevelt. In the letter Einstein urged that everything possible be done to prevent Germany from being first to develop an atomic bomb. Szilard then arranged to have the letter delivered to President Roosevelt by Alexander Sachs, a Lehman Corporation economist who had access to the White House.[5] On November 1 the President's Advisory Committee on Uranium urged that the United States initiate a crash program to study the physics of uranium fission. By early 1941 the committee had concluded that it was theoretically possible to build an atomic bomb many thousands of times more powerful than the largest bomb that had ever been made (Hewlett and Anderson 1962, p. 20; Pool 1997, pp. 31–33).[6]

Responsibility for the construction of an atomic bomb was assigned by President Roosevelt to the Army, which in turn assigned the project to the Army Corps of

4. For a useful introduction to the process of nuclear fission, see Leachman (1965).

5. For Einstein's letter and Roosevelt's reply, see Stoff, Fanton, and Williams (1991, pp. 18–20). The Einstein letter to Roosevelt did not have the immediate impact on policy that its sponsors had hoped. It was not until the end of 1941, more than two years after the letter had been presented to Roosevelt, that a decision to build the atomic bomb was finally taken. For a personal account, see Laurence (1959, pp. 37–66).

6. Institutional arrangements for providing advice to the president about issues of science and technology in military affairs evolved in two steps in the early 1940s. On June 27, 1940, President Roosevelt issued an executive order establishing the National Defense Research Council (NDRC) to be chaired by Vannevar Bush, dean of engineering at Massachusetts Institute of Technology and chairman of the National Advisory Committee for Aeronautics. In June 1941 Roosevelt issued a second executive order, establishing an Office of Scientific Research and Development (OSRD), also to be headed by Bush. The OSRD had substantially greater resources and authority than the NDRC. Bush and the OSRD played a key role in initiating the institutional arrangements for the development of both radar and the atomic bomb. The mobilization of these and other wartime projects set the stage for a prominent role for physicists in science and technology policy during World War II and during the initial years of the cold war (Dawson 1976, pp. 11–13; Kevles 1979, pp. 287–348).

Engineers. The Corps was the only agency in the federal government, with the possible exception of the Tennessee Valley Authority, that had sufficient large-scale construction experience to undertake the project. In June 1942 the Corps of Engineers formed a special district, the Manhattan District, to oversee and construct what would come to be called the *Manhattan Project*. It soon became apparent that lines of authority between the scientific community, represented by Vannevar Bush, who directed the President's Office of Scientific Research and Development, and the Army would have to be clarified. General Brehon Sommerville, commanding general of the Army Services of Supply (in which the Army Corps of Engineers was then located) selected Colonel (later Brigadier General) Leslie Groves, deputy chief of construction in the Corps, to direct the Manhattan Project. Groves brought the energy, direction, and decisiveness to the project that was necessary to assure its success.

The task that faced Groves was not only to find a workable design for the bomb, but also to create an entirely new scientific, technical, and industrial infrastructure for its production and testing. It would require both the advancement of knowledge and a unique fusion of scientific, engineering, and technical knowledge. By the time the bomb was completed and tested on July 1945, the complex included a series of university-based scientific laboratories, plutonium production facilities, uranium separation laboratories, and test reactors. Three entirely new cities—Oak Ridge, Tennessee; Los Alamos, New Mexico; and Hanford, Washington—were constructed to support the project (figure 4.1).

Research on gaseous diffusion initially was conducted at Columbia University; research on the electromagnetic separation process, at the University of California Radiation Laboratory. Development and production facilities were constructed at Oak Ridge for uranium enrichment, and at Hanford for plutonium production. The Los Alamos facility, directed by Robert Oppenheimer, was given responsibility for the actual design and production of the uranium and plutonium bombs.[7] "The manufacturing complex created by the Manhattan Project was approximately the same size as the U.S. automobile industry at that time" (Pool 1997, p. 40).

The first atomic test explosion of a plutonium bomb (designated *Trinity* by Oppenheimer) was conducted at Jornada del Muerto, New Mexico, on July 16,

7. When the Manhattan Project began, it was not possible to predict whether a uranium bomb or a plutonium bomb would be faster to build. In view of this uncertainty it was decided to try both. The Oak Ridge facility would focus on separating fissionable U-235 from natural uranium. The Hanford facilities would focus on plutonium production. Plutonium was a recently discovered new element not found in nature. It is obtained by bombarding uranium with deuterons—the nuclei of heavy hydrogen atoms. The potential advantage of plutonium was that a fissionable substance could be obtained without building isotope separation plants (Hewlett and Anderson 1962, pp. 22–23, 88–91, 308–310).

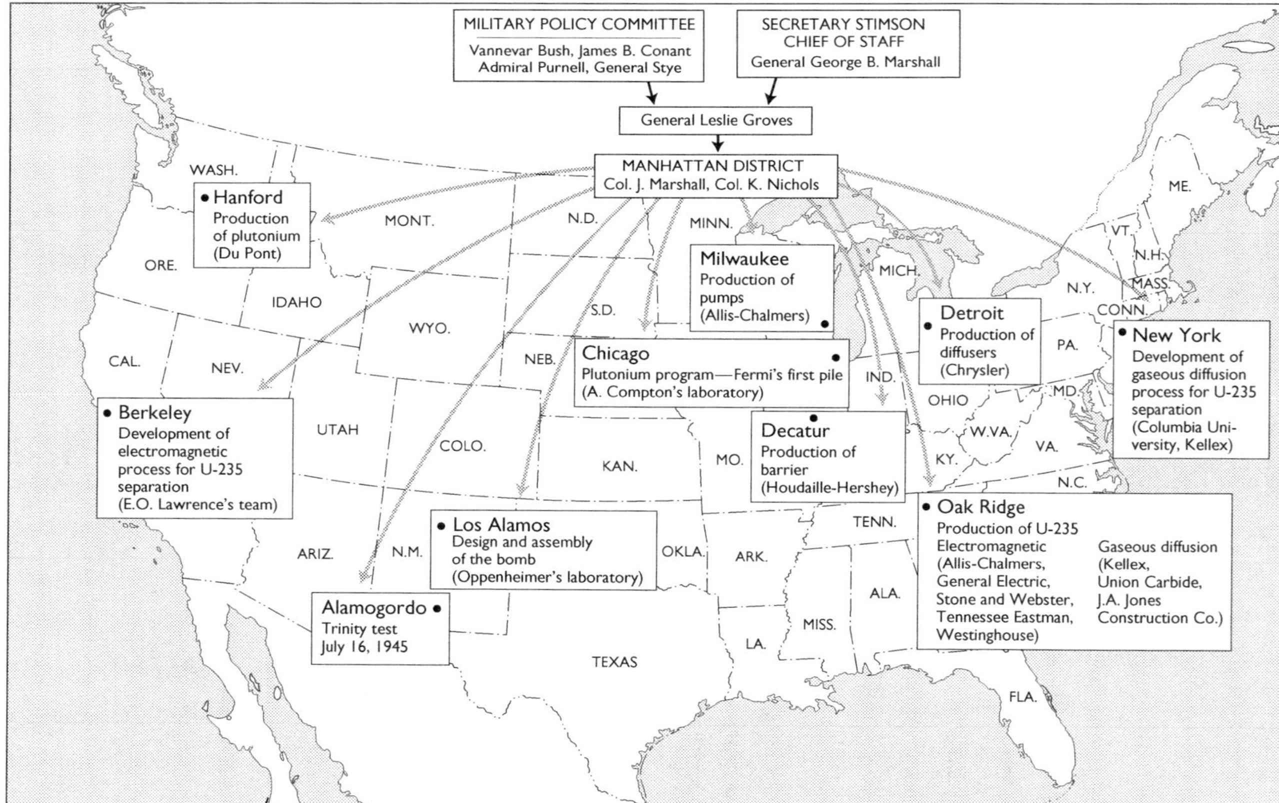

Figure 4.1. Major U.S. atomic energy facilities, 1942–1946. *Source:* Stéphane Groueff, 1967, *Manhattan Project: The Untold Story of the Making of the Atomic Bomb* (Boston: Little, Brown), frontispiece. Reprinted with permission of the author.

1945. The first uranium bomb, *Little Boy*, was detonated over Hiroshima on August 6, 1945, and the plutonium bomb *Fat Man* was detonated over Nagasaki on August 9, 1945.[8]

In retrospect it seems evident that Germany did not have the scientific, technical, and financial resources to construct an atomic bomb: "The German effort gained momentum slowly, made no real headway in solving the technical problems of building or testing a bomb, and largely petered out by the end of 1943" (Stoff, Fanton, and Williams 1991, p. 16). It has also been asserted that the Japanese government had decided to surrender before the atomic bombs were dropped on Hiroshima and Nagasaki: "Japan would have surrendered even if the atomic bombs had not been dropped, even if Russia had not entered the war, and even if no invasion had been planned or contemplated," according to the *U.S. Strategic Bombing Survey Report 1945* (quoted in Stoff, Fanton, and Williams 1991, p. 272). It is possible that both statements are correct, but this information was not available at the time the decision to drop the bombs was made (Maddox 1995).

In 1942 it had been estimated that the development of the atomic bomb would cost in the neighborhood of $100 million. In the three years between September 17, 1942, when Groves was appointed to head the Manhattan Project, and July 16, 1945, when the test explosion was conducted at the Trinity site, the total cost of the Manhattan Project had risen to approximately $2 billion (approximately $20 billion in year 2000 dollars). It is hard to believe the U.S. political system would have supported the mobilization of resources of anything like this magnitude except during war or threat of war.

Atoms for Peace

In 1946 authority to promote and regulate the development of nuclear technology for both military and nonmilitary purposes was transferred to the newly established U.S. Atomic Energy Commission (AEC). The laboratories and other facilities that had been initially established to support the work of the Manhattan Project were placed under the jurisdiction of the commission.

Initially, neither the AEC nor the power industry evidenced a great deal of enthusiasm about the prospect for nuclear power development. The AEC focused much of the effort of its laboratory system on weapons development. The private sector found the secrecy constraints imposed by the AEC cumbersome and was concerned about the long-run prospect of an adequate uranium supply: "Because

8. For a graphic description of the physical and human devastation of the bombing of Hiroshima, see Hersey (1946).

of the requirements for uranium in the weapons program dual use breeder reactors were considered necessary. The AEC initiated a uranium exploration and procurement program to alleviate this material shortage and by the early 1950s the availability of uranium appeared to no longer be a constraint" (Dawson 1976, p. 234).[9]

President Dwight D. Eisenhower's "Atoms for Peace" speech in December 1953 had the effect of committing the United States to a much more active commercial nuclear power program. The 1954 Atomic Energy Act provided a statutory basis for private-sector development of nuclear technology and for cooperation in the development of "peaceful uses" of nuclear technology with other countries (Hewlett and Holl 1989).[10]

In December 1954 the AEC, under pressure from Congress and the power industry, announced a Power Demonstration Reactor Program. Detroit Edison proposed a fast-breeder reactor; Yankee Atomic, a consortium of New England utilities, proposed a boiling-water reactor; a group headed by Commonwealth Edison of Chicago proposed a heavy-water reactor; Consolidated Edison of New York submitted an application to build a pressurized-water reactor; the Consumers Power District of Nebraska submitted plans for a sodium-graphite reactor.[11] By 1962 there were seven commercial nuclear power prototypes in operation in the

9. For an early cautious evaluation, see Schurr and Marschak (1950).

10. The 1954 Atomic Energy Act "represented a compromise among those in the administration, Congress and industry who preferred that private enterprise develop atomic energy, others who wanted a cooperative arrangement between government and private industry, and some who wished the industry to be nationalized. It allowed private corporations to build and own nuclear-power plants, but government continued to own and control the fuel" (Hughes 1989, p. 438). Before 1954 the debate about the future of atomic energy was largely confined to the Office of the President, the military, and the scientists who had been active in the development of the bomb. Private power interests had not yet acquired sufficient technical capacity to participate effectively in the debate. Industry representatives were, however, insistent that they did not want the AEC to become a "nuclear TVA"—owning and operating nuclear power-generating facilities. To the extent that public interests were involved, they were represented by the scientific community, whose major concern was that the development of nuclear energy for military and commercial use not be monopolized by the military (Dawson 1976, pp. 222–276).

11. "Nuclear reactors are classified by two of the materials used in their construction: the coolant used to transfer heat from the reactor core and the moderator used to control the energy level of the neutrons in the reactor core. In a light water reactor both the coolant and moderator are light water—H_2O. In a heavy water reactor both are heavy water—D_2O. In a gas graphite reactor the coolant is a gas, usually helium or carbon dioxide, the moderator is graphite" (Cowan 1990, p. 545). In 1955, at the first international conference on nuclear power, about a hundred types of reactor piles were discussed. Three years later the number was down to about twelve. When the U.S. Navy decided to produce a nuclear-powered submarine, after initial experiments by the AEC with six technical variants, two variants were considered, and after a single experiment with each, the light-water reactor was selected. This variant was intensively explored and developed in the following decades (Cowan 1990, pp. 547–548.)

United States (table 4.1). The effect of these actions was to generate considerable enthusiasm among the general public, the power industry, and Congress. It was frequently asserted by nuclear power enthusiasts that nuclear energy would make electric power so inexpensive that it would be "too cheap to meter" (Pool 1997, p. 71).

At the time the Power Demonstration Project was announced, the AEC had already made a decision to cooperate with Duquesne Light and Power and Westinghouse to build a pressurized light-water reactor at Shippingport, Pennsylvania. This decision was a direct consequence of a 1950 decision by the Navy to develop a water-cooled nuclear reactor to propel its first nuclear-powered submarine. President Eisenhower's "Atoms for Peace" speech in December 1953 committed the United States to a civilian nuclear power program. The reactor technology that was most readily available was the pressurized light-water technology that had been initially developed for use in nuclear-powered submarine and aircraft carriers.

The Navy's nuclear power program was directed by then Captain (later Admiral) Hyman Rickover, who simultaneously held responsibility for nuclear propulsion development in the U.S. Navy and headed the naval reactors program of the AEC. In 1946 the Navy Bureau of Ships assigned Rickover and a small contingent of civil and uniformed staff to Oak Ridge National Laboratory to evaluate the possible application of nuclear power for naval propulsion. It was recognized that nuclear-powered submarines would have clear advantages over diesel in terms of quieter operation, cruise range, speed, and the ability to remain submerged for a longer time. From 1946, through the launching of the Nautilus nuclear submarine in 1955, and to 1957, when the Shippingport power plant began operations, the development of nuclear power for both military and commercial use was almost completely dominated by Rickover's powerful engineering skills and personality (Hewlett and Duncan 1974; Hewlett and Holl 1989; Duncan 2001). The Shippingport reactor began producing electricity for commercial use in 1957, and by 1962 seven experimental nuclear electric power plants were in operation (see table 4.1): "By the mid-1960s experimentation over power reactors was over. The pressurized water reactor, by Westinghouse, and the boiling water reactor, built by General Electric, became the industry standards" (Cantelon, Hewlett, and Williams 1991, p. 305).

In the United States and later in Germany and Japan, large public research and development (R&D) programs were complemented by substantial private research investment by firms such as Westinghouse, General Electric, Babcock and Wilcox, Siemens, AEG, and Mitsubishi. In the United Kingdom, France, and the Soviet Union (Union of Soviet Socialist Republics), the research was conducted almost exclusively by the public sector. Nowhere were electric utility firms heavily involved in nuclear research. They assumed that replacing a fossil fuel–fired boiler

Table 4.1. U.S. Nuclear Electric Power Plant Prototypes in 1962

Name and Owner	Location	Type	Power Plant Reactor kw[a]	Power Plant Reactor net kw[b]	Start-Up Date
Shippingport Atomic Power Station (AEC and Duquesne Light Company)	Shippingport, PA	Pressurized water	60,000	231,000	1957
Dresden Nuclear Power Station (Commonwealth Edison Company)	Morris, IL	Boiling water	208,000	700,000	1959
Yankee Nuclear Power Station (Yankee Atomic Electric Company)	Rowe, MA	Pressurized water	161,000	540,000	1960
Indian Point Unit No. 1 (Consolidated Edison Company of New York)	Indian Point, NY	Pressurized water	255,000	585,000	1962
Hallam Nuclear Power Facility, Sheldon Station (AEC and Consumers Public Power District)	Hallam, NE	Sodium-graphite	75,000	240,000	1962
Big Rock Nuclear Power Plant (Consumers Power Company)	Big Rock Point, MI	Boiling water	47,800	157,000	1962
Elk River Reactor (AEC and Rural Cooperative Power Association)	Elk River, MN	Boiling water	20,000	58,200	1962

Source: "Nuclear Reactor," 1974, in *The New Encyclopaedia Britannica*, vol. 13: 318. Reprinted with permission of Encyclopaedia Britannica. Copyright © 1974 by Encyclopaedia Britannica.

[a] Electric output

[b] Thermal output

with a nuclear reactor to produce steam would be a relatively simple process: a nuclear reactor was just another way to boil water.

A number of different reactor designs had been advanced in the late 1950s and early 1960s. As already noted, the United States elected to use the light-water cooling and enriched uranium fuel technologies. The British and French initially used a gas graphite reactor. Canada used heavy water and natural uranium. By the mid-1960s all of the major industrial countries—the United States, Canada, the United Kingdom, the Soviet Union, France, Germany, Sweden, and Japan—were making significant investments in nuclear power generation. Improvements in reactor design and construction experience had locked the industry into the light-water, enriched-uranium path of nuclear energy development. Whether other designs would in fact have been superior in the long run is open to question, although some of the engineering literature suggests that high-temperature, gas-cooled reactors would have been superior.

Brian Arthur (1990, p. 99) and Robin Cowan (1990, pp. 541–567) have interpreted this history as an example of politically inspired "path dependence." By the time other technologies became technically and economically viable, it was too late. The path dependence was forced by strategic rather than technical or economic considerations. Without the uranium enrichment facilities built by the AEC for weapons purposes, the commercial reactors built at least through the mid-1970s would not have been economically feasible (Dawson 1976, p. 268). And Pool has insisted that without an atomic weapons program no country would have built enriched-uranium facilities (Pool 1997, p. 43).[12]

Cost Inflation

As late as the mid-1970s, the U.S. nuclear power industry seemed poised for even more rapid expansion. Restrictions by the AEC on ownership of nuclear fuel had been relaxed since the mid-1960s. By 1975 government ownership of uranium enrichment facilities was the only major exception to private ownership of the nuclear energy supply system.[13] A petroleum supply crisis that began in the early

---

12. From the late 1960s until the early 1980s, the AEC devoted very large resources to development of the liquid metal fast breeder reactor. The technical argument for the breeder reactor was its potential ability to produce more fuel than it consumed. By 1980 it became clear that the prospects for developing a commercially viable breeder would not be realized until 2025 or beyond. In 1984 appropriations for the Clinch River demonstration project were discontinued (Cohen and Noll 1991, pp. 217–257).

13. Government retention of uranium enrichment facilities was justified by the continuing concern over the ownership of the natural uranium essential to the military weapons program (Dawson 1976, p. 259).

1970s was expected to increase demand for nuclear power. It was completely unanticipated that a combination of public safety, health, and environmental concerns would bring expansion to a halt by the end of the decade.

Cost estimates by the AEC in the 1960s indicated that nuclear power capital costs would be substantially greater than those of electricity generated by large coal-fired plants. It was expected, however, that this would be compensated for by low operating costs due to the limited quantities of uranium fuel required (Weinberg 1994, p. 28). However, the anticipated economies of scale and cost reductions from "learning by doing" and "learning by using" were not realized. They were more than offset by increases in the complexity of reactors, due partly to initial design errors but largely to increasingly stringent safety standards. In many cases, final costs exceeded initial estimates by over 100 percent. It became apparent by the mid-1970s that the simple and comparatively inexpensive light-water reactors of the late 1960s were, partly on engineering grounds and partly due to safety concerns, no longer commercially viable (MacKerron 1994).

Since the early 1970s, safety requirements for nuclear plants in the United States have been continually tightened by the Nuclear Regulatory Commission in response to public risk perception. Although it is not clear that the changes in these requirements resulted in substantial safety improvements, the frequent design changes in the course of construction did result in higher construction costs. Average construction time in the United States rose to twelve years. The costs of nuclear plants of comparable size, corrected for inflation, quadrupled in little more than a decade. These higher capital costs pushed the cost of producing electricity from nuclear-fueled plants even higher as compared to coal-burning plants. In the United States no new nuclear power plants have been ordered since 1978. Plants ordered after 1974 were canceled. Strategic considerations continued to weigh heavily in decisions to sustain or expand nuclear power capacity.[14] In many developing countries nuclear power programs have absorbed a far larger share of public resources than could be justified in terms of any potential economic benefits (Marcus 1992, pp. 394–395; Abelson 1996, pp. 463–465; Solingen 1996, p. 188).

Since the late 1990s operational experience and advances in reactor technologies have led to renewed interest in the role of nuclear power in meeting future electric power demand (Taylor 2004). This economic interest has been reinforced by the potential role that nuclear power might play in reducing greenhouse gas emissions. The authors of an MIT Nuclear Energy Study (2003) argue that a nuclear energy option should be included, along with increased efficiency in electricity generation

14. In spite of the fact that no new plants were being constructed, nuclear power production in the United States increased by about 40 percent between 1990 and 2000. This increase was largely the result of improvements in plant operation (Meserve 2002).

and use, and renewable energy sources, in any comprehensive effort to reduce carbon dioxide emissions from energy generation. They argue that if nuclear energy is to play an important role in meeting the rapidly growing global demand for electricity, several unresolved problems, in addition to cost, must be successfully addressed. These are concerns about safety and health, proliferation of nuclear weapons, and disposal of nuclear wastes. The MIT study group is cautiously optimistic that these concerns can be successfully resolved over the next several decades.[15]

Power based on nuclear fission is still viewed by many scientists and engineers as a potentially environmentally benign technology capable of replacing fossil fuels on a large-scale basis (Rhoades and Beller 2000). Physicists and engineers continue to be intrigued by the possibility of producing controlled fusion reactions in a power plant to capture the large amounts of energy that are theoretically available. Fusion has two potentially important advantages. The first is that the fuel, hydrogen and its isotopes, is much less expensive and more abundant than the heavy metals, such as uranium, used in fission. The second is that, although the fusion process would create some radioactive waste, due to irradiation of the plant construction materials, it would not generate the huge amount of waste produced by fission. Fusion's major disadvantage, even if it becomes technically feasible, is one it shares with existing nuclear fission plants: the high capital cost of initial investment, which would remain an obstacle to commercial viability (box 4.1).[16]

Alternative Energy

Concern about the environmental and health implications of fossil fuel and nuclear technology, combined with the oil price shocks of the 1970s, induced an intense debate about energy futures. From the end of World War II through the early 1970s, U.S. energy R&D had focused almost entirely on nuclear energy. In 1973, for example, 67 percent of federal energy R&D expenditures were on nuclear power. Smaller amounts were spent on coal, petroleum, and natural gas. Renewable energy sources and conservation were largely ignored (Tilton 1974, pp. 8–15).[17]

By the late 1970s it was widely assumed that energy conservation could sub-

15. For a less optimistic perspective, see the review of recent literature on nuclear proliferation by Wolfsthal (2005).

16. In November 2003 the U.S. Department of Energy announced that a $5 billion contribution to the International Thermonuclear Experimental Reactor (ITER) fusion energy project ranked first among its list of scientific priorities (Malakoff and Cho 2003, pp. 1126–1127).

17. Hirsh (1999), pp. 72–117. I discuss the issues of alternative renewable and nonrenewable energy options in greater detail in Ruttan (2001, pp. 270–279).

Box 4.1. The National Energy Laboratories

The national energy laboratories operated by the U.S. Department of Energy (DOE) are perhaps the least-understood components of the U.S. national innovation system (Crow and Bozeman 1998). The 1946 legislation that established the Atomic Energy Commission (AEC) transferred to the commission the plants, laboratories, equipment, and personnel that had been assembled to build the atomic bomb.[a] A major consideration in the establishment of the AEC was to avoid military control of nuclear technology (Hewlett and Anderson 1962, pp. 1–8).

At the time the AEC was established, it "was intended to serve two main purposes: provision of large scale equipment for basic research and secure facilities for developing technologies for national security" (Westwick 2003, p. 8). It was given responsibility for governing the use of radioactive materials and for the development of nuclear technology for both military and civilian use. It maintained and expanded many of the weapons laboratories that it had inherited from the Manhattan Project, including the major multiprogram national laboratories.[b] Each of these facilities was among the largest scientific research facilities in the world. The AEC employed thousands of scientists and engineers across many disciplines, focusing chiefly on designing, testing, and manufacturing nuclear weapons systems. It provided the military with laboratory access and services (Crow and Bozeman 1998, pp. 54–55).

In the early 1950s the AEC and the laboratories were confronted with major administrative problems, such as how to maintain sufficient program autonomy to assure scientific viability and sufficient secrecy to meet the national security mission.[c] During the 1950s laboratory budgets expanded rapidly as the laboratories acquired the increasingly expensive equipment necessary to conduct basic research in subatomic and high-energy physics (Seidel 1986). Some of these developments were controversial even within the laboratory system. In 1961 Alvin Weinberg, director of Oak Ridge National Laboratory, raised three questions: "First, is Big Science ruining science?; second, is Big Science ruining us financially?; and third, should we divert a larger part of our effort toward scientific issues which bear more directly on human well being?" (p. 161).

By the early 1970s events had conspired to force a number of Weinberg's concerns onto the AEC agenda. These included the budgetary pressures associated with the Vietnam War, the slowing of productivity growth in the U.S. economy beginning in the late 1960s, and the energy shock of the early 1970s: "Even at the birthplace of AEC high energy physics, Lawrence Radiation Laboratory (LRC), scientists adapted to the new realities. New divisions—Energy and Environment, Earth Sciences, Materials, and Molecular Research emerged in the late 1960s and early 1970s to spur the laboratory to renewed growth" (Seidel 1986, p. 174). Oak Ridge National Laboratory expanded its large-scale

(*continued*)

biology program and initiated new programs, drawing on external resources, in desalinization, civil defense, natural resources, and alternative-energy research (Teich and Lambright 1976).

In 1975 the AEC was disbanded, and its staff, laboratories, and other facilities were transferred to the newly formed Energy Research and Development Administration, which was in turn consolidated into a new Department of Energy.[d] As the cold war was winding down in the 1980s, the research effort of the DOE national laboratories embraced four broad missions: (1) a *national security mission* with a primary goal of maintaining the reliability and safety of the nation's nuclear deterrent, and a secondary goal of reducing the risk of nuclear proliferation; (2) a *science mission* that provides universities and industry with world-class, large-scale scientific facilities (such as synchrotron light sources, neutron sources, and particle accelerators), supports the nation's largest federally funded research programs in the physical sciences, and contributes importantly to national programs in the environmental sciences, life sciences, and mathematics and computing; (3) an *energy mission* directed to the development of new technologies to produce energy that are affordable, environmentally acceptable, and secure, with the objectives of reducing dependence on the Persian Gulf region for energy supplies and reducing the risk of climate change associated with carbon-based energy sources; and (4) an *environmental mission* including actions to clean up the DOE nuclear weapons legacy; to stabilize, safely store, or dispose of nuclear waste; to deactivate, decontaminate, and decommission support facilities; and to remediate environmental contamination resulting from the nuclear weapons and energy programs.

During the 1980s an economic growth mission was superimposed on the other missions of the DOE national energy laboratories and on all other federally funded laboratories. It was implemented by a series of institutional innovations governing property rights in the new knowledge and technology generated by federally funded research and development (R&D; Ruttan 2001, pp. 576–581). The 1980 Stevenson-Wydler Technology Innovation Act made technology transfer a mission of all federal laboratories. The Bayh-Dole Act, passed the same year, gave title to inventions resulting from federal funding to the performers of the R&D. The Federal Technology Transfer Act of 1986 gave incentives to government-owned and-operated laboratories to commercialize their inventions, and the National Competitiveness Technology Transfer Act of 1989 extended similar rules to government-owned and contractor-operated laboratories. The latter two acts encouraged the laboratories to enter into cooperative R&D agreements (CRADAs) with industrial partners.

Initially there was substantial skepticism on the part of many students of science and technology policy as to the effectiveness of these institutional inno-

(continued)

Box 4.1. (*continued*)

vations (U.S. Secretary of Energy Advisory Board 1994). Over time, however, careful empirical studies have demonstrated that CRADAs have been much more effective than anticipated in stimulating both industrial patents and company-financed R&D (Cohen and Noll 1996; Jaffe and Lerner 2001; Adams, Chiang, and Jensen 2003). In spite of this accumulating evidence, a sharp reaction had emerged by the mid-1990s against the dual-use and cooperative programs that had been directed to enhancing technology transfer from the national laboratories to the private sector. The argument was not that the programs were ineffective, but that in an era of budget stringency they were diverting effort from traditional defense and energy-related missions (U.S. Secretary of Energy Advisory Board, 1994; Bozeman and Dietz 2001).

It is doubtful that the new industrial innovation policies introduced in the 1980s will lead to the development of new general-purpose technologies. They have generated incremental rather than radical technical innovations. They have had great difficulty in achieving institutional viability within the national energy laboratory system and in gaining sustained political support. It is possible, though unlikely, that a national commitment to the development of alternatives to carbon-based energy sources could focus the R&D of the national energy laboratory system on development of new environmentally compatible general-purpose technologies. (I return to this issue in chapter 8.)

Sources: In preparing this box I found Crow and Bozeman (1998) and Westwick (2003) particularly useful. In addition to works cited in the text, I draw on Curtis, McTeague, and Cheney (1997) and Lawler (1996).

[a] A major consideration in the establishment of the AEC was to avoid military control of nuclear technology. The AEC was governed by a five-man board. The board was chaired by David Lilienthal, formerly chairman of the Tennessee Valley Authority.

[b] Initially the term *national laboratory* was reserved for a limited number of multiprogram laboratories that were engaged in basic research, such as Argonne, Berkeley, Brookhaven, Los Alamos, and Oak Ridge. Other sites, such as Hanford and Sandia, that were initially focused on the production of nuclear weapons and weapons material were not covered by the term. As their scope of research broadened over time, the term became more inclusive (Westwick 2003, p. 9).

[c] For a very thorough discussion of the program, administrative, and political issues that confronted the national energy laboratory system between the late 1940s and the early 1970s, see Westwick (2003).

[d] In the two decades after the incorporation of the AEC into the DOE, more than twenty major commissions or task forces were chartered to address the question of just what the United States should expect from its system of national energy laboratories. For a critique of these efforts, see Crow and Bozeman (1998).

stantially slow the rate of growth in energy use and that by the end of the century renewable energy sources could account for a substantial share of incremental growth in energy production. It was recognized, however, that the substitution of renewable energy for fossil fuel and nuclear sources, and the slowing of energy use, could be achieved only by changes in the technology of electric power generation, in the technology of energy use, and in the incentives facing both producers and consumers.

A series of government interventions, beginning with the Clean Air Act of 1970 and the Public Utility Regulatory Policies Act of 1978, were designed to address the environmental health implications of electric power production. There have also been interventions designed to encourage the development and adoption of energy-conserving technologies and practices. By the mid-1990s it was clear that the changes in technology and policy were quite different from the changes that had been anticipated by reform advocates in the 1970s and 1980s. Design improvements have led to rapid improvements in the efficiency of gas turbines for the generation of electricity and to rapid increase in the use of natural gas as a primary energy source for the production of electricity (Martin 1996; Alic, Mowery, and Rubin 2003, pp. 13–14). The changes in energy technology and policy since the early 1970s, particularly the substitution of natural gas for coal in electricity generation, have been associated with continuation of the long-term trend toward energy sources with lower carbon content.[18]

If the trend toward decarburization is to continue into the middle of the twenty-first century and beyond, it may depend on the use of pure hydrogen as a fuel. This will require the development of an economically viable technology for electrolyzing water (Ausubel 1991; Grübler and Nakicenovic 1996). Water (H_2O) can be split into hydrogen and oxygen by having an electric current passed through it. In the 1950s and 1960s, it was anticipated that the cost of electricity produced by nuclear power would be low enough to make electrolytic hydrogen production economically viable. In the 1980s and the 1990s, advances in photoelectric cell technology again created considerable optimism that hydrogen would become an economically viable fuel (Ogden and Williams 1989). The Bush administration announced in January 2003 a commitment to a $1.2 billion research initiative to replace carbon with hydrogen-based fuels in the field of transportation. The prospect that a "hydrogen economy" will be successfully implemented during the first

18. As this book was being completed, price increases were slowing the transition to use of natural gas in electric power production. By the mid-2000s coal-fired electricity-generating plants were accounting for an increasing share of new generating capacity in the United States and in several rapidly growing developing countries such as China and India (Barta and Smith 2004).

half of this century, however, remains controversial (National Research Council and National Academy of Engineering 2004; Romm 2004; Sperling and Ogden 2004).

Perspective

Nuclear power is the most clear-cut example discussed in this book of an important general-purpose technology that in the absence of military and defense-related procurement would not have been developed at all—it would not have been developed "anyway." It is exceedingly difficult to imagine that, without the threat of Germany's developing nuclear weapons during World War II, the U.S. government would have mobilized the scientific, technical, and financial resources devoted to the Manhattan Project. It is equally difficult to imagine circumstances other than the cold war that would have enabled the U.S. government to sustain its investment in nuclear energy into the 1980s.

What if there had been no Manhattan Project? Pool has argued that, in the absence of an atomic weapons program, the United States would not have built nuclear enrichment facilities. And without the enriched uranium supplied by the AEC's weapons program, it is unlikely that a nuclear navy program would have been implemented or that a nuclear power program would have been developed (Pool 1997, p. 43). Chauncey Starr, one of the more experienced and thoughtful observers of the nuclear power industry, speculated in the mid-1990s that, in the absence of the threat of war, Hahn and Strassman's work would have been written up in the scientific literature and treated as a subject of mostly academic interest. Because of the high cost, research and development of atomic reactors would have proceeded at a modest pace. Low-power nuclear reactors would have been developed to produce isotopes primarily for medical and industrial applications (Pool 1997, p. 41).

During the last quarter of the nineteenth century, the electric light and power industry emerged as the most dynamic general-purpose technology in the U.S. economy. During the first half of the twentieth century, it was a major source of productivity and economic growth in the United States and other industrial economies (Ruttan 2001, pp. 247–266). At mid-century there were great expectations that the exploitation of nuclear energy would enable the electric power industry to renew its role as a dynamic source of economic growth by making electricity available to industry and consumers on increasingly favorable terms—"too cheap to meter." These expectations have not been realized.

It is possible that during the first half of the twenty-first century nuclear power will be able to make a significant contribution to meeting the growth in demand for electric power, and by substituting for carbon-based fuels, to contribute to slowing the accumulation of greenhouse gases in the atmosphere. In retrospect it seems quite possible that if the United States had proceeded at a more measured pace in the development and introduction of nuclear power in the 1950s and 1960s, it would today be in a stronger position to bring nuclear technology to bear in meeting the demands of the twenty-first century.

ACKNOWLEDGMENTS

I am indebted to Richard Fuller, Robert Seidel, and Douglas Tiffany for comments on an earlier version of this chapter.

REFERENCES

Abelson, P. H. 1996. "Nuclear Power in East Asia." *Science* 272: 165.

Adams, J. A., E. P. Chiang, and J. E. Jensen. 2003. "The Influence of Federal Laboratory R&D on Industrial Research." *Review of Economics and Statistics* 85: 103–120.

Alic, J. A., D. C. Mowery, and E. S. Rubin. 2003. *U.S. Technology and Innovation Policies: Lessons for Climate Change.* Arlington, VA: Pew Center for Global Climate Change.

Arthur, W. B. 1990. "Positive Feedbacks in the Economy." *Scientific American* 262 (no. 2): 92–99.

Ausubel, J. H. 1991. "Energy and the Environment: The Light Path." *Energy Systems and Policy* 15: 181–188.

Barta, P., and R. Smith. 2004. "Global Surge in Use of Coal Alters Energy Equation." *Wall Street Journal,* 10 November, A-1, A-17.

Bozeman, B., and J. S. Dietz. 2001. "Research Policy Trends in the United States: Civilian Technology Programs, Defense Technology and the Deployment of the National Laboratories." In *Research and Innovation Policies in the New Global Economy: An International Comparative Analysis,* ed. P. Laredo and P. Mustar. Cheltenham, U.K.: Elgar.

Cantelon, P. L., R. G. Hewlett, and R. C. Williams, eds. 1991. *The American Atom: A Documentary History of Nuclear Policies from the Discovery of Fission to the Present.* Philadelphia: University of Pennsylvania Press.

Cohen, L. R., and R. G. Noll. 1991. "The Clinch River Breeder Reactor." In *The Technology Pork Barrel,* ed. L. R. Cohen and R. G. Noll. Washington, DC: Brookings Institution Press, 217–258.

———. 1996. "The Future of the National Laboratories. *Proceedings of the National Academy of Sciences* 93: 12678–12685.

Cowan, R. 1990. "Nuclear Power Reactors: A Study in Technological Lock-In." *Journal of Economic History* 50: 541–567.

Crow, M., and B. Bozeman. 1998. *Limited by Design: R&D Laboratories in the U.S. National Innovation System.* New York: Columbia University Press.

Curtis, C. B., J. M. McTeague, and D. W. Cheney. 1997. "Fixing the National Laboratory System." *Issues in Science and Technology* (Spring), 49–56.

Danhof, C. H. 1968. *Government Contracting and Technological Change.* Washington, DC: Brookings Institution.

Dawson, F. G. 1976. *Nuclear Power: Development and Management of a Technology.* Seattle: University of Washington Press.

Duncan, F. 2001. *Rickover: The Struggle for Excellence.* Annapolis, MD: Naval Institute Press.

Groueff, S. 1967. *Manhattan Project: The Untold Story of the Making of the Atomic Bomb.* Boston: Little, Brown.

Grübler, A. 1998. *Technology and Global Change.* Cambridge: Cambridge University Press.

Grübler, A., and N. Nakicenovic. 1996. "Decarbonizing the Global Energy System." *Technological Forecasting and Social Change* 53: 97–110.

Hersey, J. 1946. *Hiroshima.* New York: Knopf.

Hewlett, R. G., and O. E. Anderson, Jr. 1962. *A History of the United States Atomic Energy Commission, Vol. 1: The New World, 1939/1946.* University Park: Pennsylvania State University Press.

Hewlett, R. G., and F. Duncan. 1962. *A History of the United States Atomic Energy Commission, Vol. 2: Atomic Shield: 1947/1947.* University Park: Pennsylvania State University Press.

———. 1974. *Nuclear Navy: 1946–1962.* Chicago: University of Chicago Press.

Hewlett, R. G., and J. M. Holl. 1989. *Atoms for Peace and War: 1953–1961.* Berkeley: University of California Press.

Hirsh, R. F. 1999. *Power Loss: The Origins of Deregulation in the Restructuring of the American Electric Utility System.* Cambridge, MA: MIT Press.

Hughes, T. P. 1983. *Networks of Power: Electrification in Western Society, 1889–1930.* Baltimore: Johns Hopkins University Press.

———. 1989. *American Genesis: A Century of Invention and Technological Enthusiasm, 1870–1970.* New York: Viking.

Jaffe, A. B., and J. Lerner. 2001. "Reinventing Public R&D: Patent Policy and the Commercialization of National Laboratory Technologies." *RAND Journal of Economics* 32: 167–198.

Kevles, D. J. 1979. *The Physicists: The History of a Scientific Community in Modern America.* New York: Random House.

Laurence, W. L. 1959. *Men and Atoms: The Discovery, the Uses, and the Future of Atomic Energy.* New York: Simon & Schuster.

Lawler, Andrew. 1996. "DOE Labs: Is Evolution Enough?" *Science* 272: 1576–1578.

Leachman, H. B. 1965. "Nuclear Fission." *Scientific American* 213 (August): 49–59.

MacKerron, G. 1994. "Innovation in Energy Supply: The Case of Electricity." In *The Handbook of Industrial Innovation,* ed. M. Dodgson and R. Rothwell. Aldershot, U.K.: Elgar, 182–190.

Maddox, R. J. 1995. *Weapons for Victory: The Hiroshima Decision Fifty Years Later.* Columbia: University of Missouri Press.

Malakoff, D., and A. Cho. 2003. "ITER Tops DOE List of Next Big Science Projects." *Science* 302: 1126–1127.

Marcus, A. A. 1992. *Controversial Issues in Energy Policy*. Newbury Park, CA: Sage Publications.

Martin, J. M. 1996. "Energy Technologies: Systemic Aspects, Technological Trajectories, and Institutional Frameworks." *Technological Forecasting and Social Change* 53: 81–95.

Meserve, R. 2002. "The Changing Climate for Nuclear Power in the United States." *American Academy of Arts and Sciences Bulletin* (Winter), 57–72.

MIT Nuclear Energy Study. 2003. *The Future of Nuclear Power*. Cambridge: Massachusetts Institute of Technology.

National Resource Council and National Academy of Engineering. 2004. *The Hydrogen Economy: Opportunities, Costs, Barriers, and R&D Needs*. Washington, DC: National Academies Press.

"Nuclear Fission, Nuclear Fusion, Nuclear Reactors, and Nuclear Weapons." 1974. In *Encyclopaedia Britannica*, vol. 13: 301–328.

Ogden, J. M., and R. H. Williams. 1989. *Solar Hydrogen Moving beyond Fossil Fuels*. Washington, DC: World Resources Institute.

Pool, R. 1997. *Beyond Engineering: How Society Shapes Technology*. New York: Oxford University Press.

Rhoades, R., and D. Beller. 2000. "The Need for Nuclear Power." *Foreign Affairs* 79: 30–44.

Romm, J. J. 2004. "The Hype about Hydrogen." *Issues in Science and Technology* 20 (Spring): 74–81.

Ruttan, V. W. 2001. *Technology, Growth, and Development: An Induced Innovation Perspective*. New York: Oxford University Press.

Schurr, S. C., and J. Marschak. 1950. *Economic Aspects of Atomic Power*. Princeton, NJ: Princeton University Press.

Seidel, R. W. 1986. "A Home for Big Science: The Atomic Energy Commission Laboratory System." *Historical Studies in the Physical Sciences* 18: 135–176.

Solingen, E. 1996. *Industrial Policy, Technology and International Bargaining: Designing Nuclear Industries in Argentina and Brazil*. Stanford, CA: Stanford University Press.

Sperling, D., and J. Ogden. 2004. "The Hope for Hydrogen." *Issues in Science and Technology* 20 (Spring): 82–86.

Stoff, M. B., J. F. Fanton, and R. H. Williams. 1991. *The Manhattan Project: A Documentary Introduction to the Atomic Age*. Philadelphia: Temple University Press.

Taylor, J. J. 2004. "The Nuclear Power Bargain." *Issues in Science and Technology* 20 (Spring): 41–47.

Teich, A. H., and H. Lambright. 1976. "The Redirection of a Large National Laboratory." *Minerva* 14: 447–474.

Tilton, J. E. 1974. *U.S. Energy R&D Policy*. Washington, DC: Resources for the Future.

U.S. Secretary of Energy Advisory Board. 1994. *Task Force on Alternative Futures for the Department of Energy National Laboratories*. Washington, DC: U.S. Government Printing Office (Galvin Commission Report).

Weinberg, A. M. 1961. "Impact of Large-Scale Science in the United States." *Science* 134: 161–164.

———. 1994. *The First Nuclear Era: The Life and Times of a Technological Fixer*. New York: American Institute of Physics Press.

Westwick, P. J. 2003. *The National Labs: Science in an American System, 1947–1974*. Cambridge, MA: Harvard University Press.

Wolfsthal, J. B. 2005. "The Next Nuclear War: Non-Proliferation in a New World." *Foreign Affairs* 84 (February): 156–161.

5

The Computer Industry

The development of the electronic digital computer was preceded by a long history of mechanical and electromechanical tabulating machines. In the latter years of the nineteenth century, numerous office machines—typewriters, adding machines, cash registers, mechanical calculators, and billing and accounting devices—were introduced.[1] In 1886 Herman Hollerith, a statistician employed at the U.S. Census Office, designed and built an electrically run tabulator that used punch card inputs and electrical card reading. The Hollerith machine was able to process census data in one third of the time it would have taken with handwritten tally sheets (Cortada 1993; Ruttan 2001, pp. 317–319).

In this chapter I trace the role of military procurement on the development of the computer industry. During and immediately after World War II, major efforts were made, with the support of the military, to develop fully electronic computing machines. The role of the military in driving the development of computer, semiconductor, and software technologies cannot be overemphasized. Until well into the 1960s, these technologies were nourished by markets that were almost completely dependent on the defense, energy, and space agencies and industries. An important subtheme in this chapter is the intimate relationship between advances in science and those in technology in the development of the computer, the transistor, and computer software.

1. In this chapter I have drawn extensively on "The Computer and Semiconductor Industries" in Ruttan (2001, pp. 316–367). I define the computer industry broadly to include semiconductors and software.

Inventing the Computer

The first fully automatic calculator was a product of collaboration between Harvard University and International Business Machines (IBM). The Automatic Sequence Controlled Calculator (Mark I), an electromechanical machine, was completed in 1944. It could add, subtract, multiply, divide, and table reference. Input data were entered on punched cards, and output was recorded either on punched cards or on an electric typewriter. Early models were built for the Navy and Air Force.[2] In 1947, drawing on its experience in constructing the Mark I, IBM constructed a "super calculator," the Selective Sequence Electronic Calculator (SSEC). It was a "gigantic hybrid of electronic and mechanical parts, half modern computer and half punch card machine" (Watson 1990, p. 190).

Firing Tables and Cryptology

The first all-purpose electronic digital computer was developed by John W. Mauchly and J. Presper Eckert and associates at the University of Pennsylvania's Moore School of Electrical Engineering.[3] The Army Ballistics Research Laboratory (BRL) was confronted with the enormous labor involved in calculating artillery firing tables. Against the advice of the U.S. computing establishment, it "chose to gamble on an untested technology" (Flamm 1988, p. 252). The Mauchly-Eckert machine, the Electronic Numerical Integrator and Calculator (ENIAC), was completed in 1946. It was capable of computing more than a thousand times faster than any available electromechanical machine: "The ignition of the hydrogen bomb

2. The best short treatment of the role of military procurement in computer development is Seidel (2002). The most useful book-length treatments of the development of the computer are Ceruzzi (2003) and Flamm (1987; 1988). In this section I also draw on Bashe et al. (1986), Katz and Phillips (1982), Pugh (1984, pp. 301–312), Shurkin (1984), Campbell-Kelley and Aspray (1996), Mowery and Rosenberg (1998, pp. 135–151), and National Research Council (1999, pp. 85–135).

3. The electronic digital computer was conceived by John V. Atanasoff of Iowa State University in 1937. In December 1940 he demonstrated a small prototype, and in 1941 he published an article on the theory and design of computers. In 1941 Mauchly visited Iowa State to examine the computer, read the technical papers, and discuss his work with Atanasoff. Iowa State failed to patent the Atanasoff design. Although the issue remains controversial, most historians of computing now credit Atanasoff rather than Mauchly and Eckert as the inventor of the electronic digital computer (Shurkin 1984, pp. 114–116; Slater 1987, pp. 53–79). For other precursors to Mauchly and Eckert in the development of the electronic digital computer, see Lee (2002).

was simulated in the first program run on the ENIAC" (Seidel 2002, p. 191).[4] The successful completion of the ENIAC provided a great impetus for the development of the computer by stimulating further defense agency demand, even though the ENIAC had no immediate commercial applications.

A second computer developed by the Moore School group, the Electronic Discreet Variable Computer (EDVAC), had an even more important impact on future computer development. It incorporated the concept of a stored program and sequential processing developed by Mauchly, Eckert, and Herman Goldstein of the Moore group and the mathematician John von Neumann of Princeton University's Institute for Advanced Study. In what came to be referred to as the *von Neumann architecture*, the processing unit of the computer fetched instructions from a central memory that stored both data and programs, operated on the data (for example, added or subtracted), and returned the results to a central memory.[5]

In the early postwar period, there was rapid formation and consolidation of firms to exploit the new technology under the impetus of defense agency demand (Norberg 2002). Eckert and Mauchly formed the Electric Control Company in June 1946 and the Eckert-Mauchly Computer Corporation (EMCC) in 1947. Because they had difficulty raising significant capital to complete their development work, Eckert and Mauchly accepted an offer to be acquired by Remington Rand in 1950. They brought with them several contracts, including one with the Bureau of the Census to develop an EDVAC-type computer, the Universal Automatic Computer (UNIVAC).

A second pioneering computer company, Engineering Research Associates (ERA), was formed in St. Paul, Minnesota, in 1946. The founders of ERA were drawn from the Naval Communications Supplemental Activity, located in St. Paul, which had been involved in developing computers in support of the Navy's work in cryptology.[6] "With ERA," summarizes the National Research Council (NRC),

4. Flamm notes that the decision was facilitated by the support of John von Neumann, "who was concerned about the enormous computational demands of nuclear weapons design" (1988, p. 252). The initial cost of building the ENIAC was estimated as $50,000. The final cost was in the neighborhood of $500,000 (Edwards 1996, p. 51).

5. The first electronic digital von Neumann–type stored-program computer to be placed in regular operation in the United States was the Standard Eastern Automatic Computer completed by the National Bureau of Standards in 1950. It was the source of a number of important technical innovations, including solid-state logic. Its development was supported by both the Navy and the Army (Flamm 1988, pp. 68–75).

6. As early as 1943 a team led by Alan Turing at the British Government Code and Cypher School developed an electronic digital computer, the Colossus, designed to automate the decoding of German military communications. The machine could not store programs internally (Edwards 1996, pp. 16–18).

"the Navy effectively privatized its wartime cryptography and was able to maintain civilian expertise through the radical postwar demobilization" (1999, p. 91). ERA's first major computer system, the Atlas, was delivered to the Navy in December 1950. A modified version designed for commercial applications, the ERA 1101, became available in 1952. ERA also ran into financial difficulties and agreed to be acquired by Remington Rand in 1952. With its Eckert-Mauchly and ERA operations, Remington Rand controlled a significant share of the total computer engineering capacity in the United States (Tomash and Cohen 1979).

Both Eckert-Mauchly and the ERA group were disappointed by Remington Rand's lack of enthusiasm for commercial development. This lack of enthusiasm was shared by other office equipment manufacturers (Cortada 1993, pp. 222–246). In 1950 IBM president Thomas Watson, Sr., reportedly asserted that the SSEC, the electromechanical machine developed by IBM and on display at the IBM headquarters in New York City, was sufficient to "solve all the important scientific problems in the world involving scientific calculations" (Katz and Phillips 1982, p. 171). Although Watson apparently saw only limited commercial possibilities for computers, IBM was already making substantial investment in electronics research.

It was the Korean War that led to a decision by IBM to test the market for commercial computers. Assured by letters of intent from government agencies and defense-related firms, IBM initiated the Defense Calculator project in early 1951. When a rental price of $15,000 per month was announced, many of the prospective clients withdrew (Katz and Phillips 1982, p. 177). IBM decided, however, to go ahead with development of the machine, to be renamed the IBM 701. At the end of March 1953, the first external installation of an IBM 701 took place at the U.S. Atomic Energy Commission's (AEC) Los Alamos Laboratory (MacKenzie 1996, p. 114). "Replacement of the SSEC as the show piece of IBM's computer capability signaled the transition of IBM to a new era of postwar electronic computer technology" (Pugh 1984, p. 32).[7]

7. Arthur Norberg has pointed out that Eckert-Mauchly (EMCC), ERA, and IBM required about the same length of time to design and deliver a computer system. EMCC began first in 1946 and delivered the Binary Automatic Computer to Northrop Aviation in 1949. ERA received a contract to design and build the Atlas in 1947 and delivered the machine to the National Security Agency in 1953. Development work on the IBM Defense Calculator and the IBM 650 project began in 1948. The defense calculator appeared in 1951; a commercial version, the IBM 701, in December 1952; and the IBM 650, in 1954. Thus, by the mid-1950s the three firms had achieved roughly comparable capacity (Norberg 1993, p. 193). For greater detail on the early history of ERA, see Tomash and Cohen (1979). For a highly personal account of the cautious entry of IBM into the electronic digital computer field, see Watson (1990, pp. 130–146, 188–207, 227–238). Cortada has emphasized that the practices worked out over the previous several decades in the office machinery business "enabled IBM management to enter the com-

Whirlwind and the Semi-Automatic Ground Environment

Intensification of the cold war in the early 1950s played a critical role in the development of IBM's capacity to market a fully transistorized commercial computer (Usselman 1993).[8] The impetus came from a decision by IBM to cooperate with the Massachusetts Institute of Technology (MIT) Lincoln Laboratory in the design and development of a computerized air defense system, the Semi-Automatic Ground Environment (SAGE), funded by the U.S. Air Force. The computer technology developed for the SAGE Project was a direct outgrowth of Project Whirlwind, originally intended to develop a general-purpose flight simulator, which had been initiated during World War II by a group of young graduate engineers led by Jay Forrester at the MIT Servomechanisms Laboratory and funded by the Office of Naval Research (ONR).

In 1946 Forester succeeded in convincing the U.S. Department of Defense (DOD) to commit the large resources needed to expand the goal of the Whirlwind program to the design of a real-time general-purpose digital computer that could serve functions other than flight simulation. In 1950, as the development of the Whirlwind computer was being completed, the ONR, facing increasing budgetary pressure and skeptical that Whirlwind would be as useful as had earlier been anticipated, entered into negotiations with MIT about reducing its level of funding for the project. A fortuitous visit to the Servomechanism Laboratory in January 1950 by George E. Valley, Jr., an MIT physics professor and chairman of the Air Defense System Engineering Committee, led to Valley's interest in the potential role that Whirlwind might play in an Air Force project to develop a computer-based air defense system. Arrangements were made to shift the major funding of Whirlwind from ONR to the Air Force.[9]

The task envisaged for SAGE was to detect alien aircraft, select appropriate

---

puter business with proven methods of marketing and support and a customer base that could migrate to the new technology" (1993, p. 127).

8. The IBM decision was also influenced by a 1951 U.S. Justice Department antitrust suit against IBM over its policy of only leasing its tabulating machines, monopolization of the punch card market, and discrimination in punch card sales. IBM president Thomas J. Watson, Jr., agreed to a consent decree resolving the antitrust suit in 1956. His strategy was to forego dominance in the mature tabulating and card markets and achieve dominance in the computer market. Air Force and AEC contracts contributed substantially to the success of Watson's strategy (Jorgensen 1996).

9. For the definitive study of the Whirlwind Project and of the role of the MIT team directed by Jay Forrester in the development of Whirlwind and SAGE, see the book *From Whirlwind to MITRE* (Redmond and Smith 2000); see also National Research Council (1999, pp. 92–95).

interceptor aircraft, and determine antiaircraft missile trajectories. Critical to the SAGE system would be two pieces of equipment that in 1950 still had to be designed, built, and tested: (1) a high-speed electronic digital processing machine that would be located at each radar site to perform real-time processing of radar signals and send the data over telephone lines to the command-and-control center, and (2) a central computer located at the command-and-control center that would accept and process the data from the distributed radar sites. The system would have to store and process large amounts of information and coordinate several computers in real time (Redmond and Smith 2000, p. 2). By 1951 work had progressed to the point where a successful simulation test involving live aircraft could be carried out over Medford, Massachusetts (Redmond and Smith 2000, p. 3).

By the middle of 1952 sufficient progress had been made in the design of Whirlwind II that the project team initiated efforts to identify a manufacturer who might have the capacity to build a prototype and a production model of the machine. Preliminary inquiry identified five firms as potential collaborators: the Bell System (Bell Telephone Laboratories and Western Electric), Radio Corporation of America (RCA), Raytheon, Remington Rand (Eckert-Mauchly and ERA), and IBM. Following meetings with representatives of each firm and visits to facilities by Whirlwind managers, Bell and RCA withdrew from further consideration because of the pressure of prior staff commitments. Raytheon was thought to lack engineering capacity, and the two divisions of Remington Rand were not sufficiently integrated. By a substantial margin, IBM was judged to be the best choice for a collaborator.

In fall 1952 managers and engineers at MIT and IBM initiated a period of intense and frequently stressful collaboration that led from the building of an air defense computer to an air defense system. In 1953 and 1954 experimental tests of the system were conducted over Cape Cod (the Cape Cod System). In 1958 the first sector of the SAGE system became fully operational. In December 15, 1961, the last North American link in the SAGE air defense system, which included twenty-one sectors stretching along the East and West Coasts, along the northern states, and into northern Canada were turned over to the Air Defense Command. By the time it was completed, IBM had built fifty-four computers for the SAGE system.

The SAGE project was a driving force behind the commercial development of the American computer industry and has been deemed the most important learning experience in computer history. It led to many of the inventions that we have come to expect in our personal computers (Katz and Phillips 1982; Hughes 1998, pp. 15–67): "It revolutionized the information industry by spanning in one inspired leap the prehistoric computer era of serial batch processing and the modern world

Table 5.1. Early U.S. Support for Computers

First Generation of U.S. Computer Projects	Estimated Cost of Each Machine (Thousands of Dollars)	Source of Funding	Initial Operation
ENIAC	750	Army	1945
Harvard Mark II (partly electromechanical)	840	Navy	1947
Eckert-Mauchly BINAC	278	Air Force (Northrop)	1949
Harvard Mark III (partly electromechanical)	1,160	Navy	1949
NBS Interim computer (SEAC)	188	Air Force	1950
ERA 1101 (Atlas I)	500	Navy/NSA	1950
Eckert-Mauchly UNIVAC	400–500	Army via Census; Air Force	1951
MIT Whirlwind	4,000–5,000	Navy; Air Force	1951
Princeton IAS computer	650	Army; Navy; RCA; AEC	1951
University of California CALDIC	95	Navy	1951
Harvard Mark IV	—	Air Force	1951
EDVAC	467	Army	1952
Raytheon Hurrican (RAYDAC)	460	Navy	1952
ORDVAC	600	Army	1952
NBS/UCLA Zephyr computer (SWAC)	400	Navy, Air Force	1952
ERA Logistics computer	350–650	Navy	1953
ERA 1102 (3 built)	1,400	Air Force	1953
ERA 1103 (Atlas II, 20 built)	895	Navy/NSA	1953
IBM Naval Ordinance Research Computer (NORC)	2,500	Navy	1955

Source: Kenneth Flamm, 1988, *Creating the Computer: Government, Industry and High Technology* (Washington, DC: Brookings Institution Press), 76.

of interactive systems" (Redmond and Smith 2000, p. 442).[10] During the first two decades of computer development, its technical cutting edge was advanced primarily by government laboratories or by private industry engaged in military and defense-related research (National Research Council 1999, pp. 86–96).

By the late 1950s progress in semiconductor development was beginning to open up the possibility of designing smaller and less expensive computers. In 1957 Kenneth Olson, a former IBM employee, and Harlan Anderson founded Digital Equipment Corporation (DEC) with $30,000 in venture capital funding. Olson was committed to a vision of computers that were smaller, easier to operate, and much less expensive than an IBM mainframe (Rifkin and Harrar 1988). The first DEC computer, the PDP-1, was demonstrated in 1959: "It sold for $120,000, contained 4K bytes of memory, was the size of a refrigerator, and included a cathode ray television-like video display built in the console" (Langlois 1992, p. 7). The PDP-8, introduced in 1965, was the first computer to use integrated circuits. The initial commercial success of the DEC computers encouraged the entry of other firms, such as Data General, Scientific Data Systems, Hewlett-Packard, and Wang, to enter the minicomputer market. From a long-term perspective, the significance of the minicomputers is that they provided a bridge between the mainframe and the microcomputer, for "the minicomputer generated the seeds of its own destruction" (Ceruzzi 2003, p. 206). The minicomputer market peaked in the early 1980s. The development of the microcomputer was dependent on the development of semiconductors—the transistor and the microprocessor.[11]

---

10. The innovations made in connection with the SAGE project at MIT and IBM included (1) techniques to manufacture ferrite core memory rapidly, inexpensively, and reliably; (2) computer-to-computer telecommunications; (3) real-time simultaneous use by many operators; (4) keyboard terminals for man-machine interaction; (5) simultaneous use of two linked computers; (6) ability to devolve certain functions to remote locations without interfering with the dual processors; (7) use of display options independent of dual processors; (8) inclusion of an interrupt system, diagnostic programming, and maintenance warning techniques; and (9) memory development (Katz and Phillips 1982, p. 185; National Research Council 1999, pp. 92–94). Most of these advances were incorporated into IBM's first fully transistorized computer, the 7090. Other IBM 7090 innovations included (1) radically new parallel architecture, permitting several operations to be performed simultaneously; (2) standard modular systems component technology; (3) printed circuit cards and improved back-wiring; (4) an 8-bit byte; (5) greatly improved transistors and the means of manufacturing them; (6) a common mode for attaching peripherals; (7) a combination of decimal and binary arithmetic; and (8) combined fixed and variable word length operations (Katz and Phillips 1982, p. 189). Pugh notes that the development of high-speed ferrite core memories that could be mass produced at low cost was probably the most important innovation that made stored-program computers a practical commercial reality (1984, p. ix).

11. Military procurement played only an indirect role in the development of the microcomputer. For an excellent review of the development of the microcomputer—the personal computer—see Ceruzzi (2003, pp. 207–241). See also Ruttan (2001, pp. 331–337).

Semiconductors

It was understood even in the 1940s that the speed, reliability, physical size, and heat-generating properties of the vacuum tubes used in telephone switching devices would become a major technical constraint on electric switching.[12] The first working transistor emerged out of research led by William Shockley of the Bell Telephone Laboratories solid-state research group in late 1947.[13] Shockley joined Bell in 1936 after receiving a Ph.D. in physics from MIT. Shortly after Shockley joined Bell Laboratories, Mervin Kelly, director of research, emphasized to Shockley his interest in developing electronic switching, in which metal contacts would be replaced by electronic devices. In the late 1930s Shockley began to consider the possibility of an approach to developing electronic switching devices, an approach based on solid-state physics.

After World War II Bell Laboratories formed a Solid State Department to develop new knowledge that might be used in the development of completely new and improved components and apparatuses for communications systems. In attempting to understand why a prototype semiconductor amplifier developed earlier by Shockley had failed, two colleagues, John Bardeen and Walter Brattain, produced the first working transistor (the point-contact design) on December 15, 1947. Their work led to an effort by Shockley to develop the bipolar junction transistor. A satisfactory design was not achieved, however, until spring 1950. Advances in process engineering, particularly the development of techniques for producing germanium and silicon crystals, were required before production of the junction transistor would become feasible (Shockley 1976, pp. 597–620; Teal 1976, pp. 621–639; Bassett 2002, pp. 12–22).[14]

12. In this section I draw heavily on Nelson (1962); Katz and Phillips (1982); Levine (1982); Mowery (1983, pp. 183–197); Riordan and Hoddeson (1997); and Mowery and Rosenberg (1998, pp. 124–135, 151–152). For discussion of the scientific and engineering aspects of semiconductors, including circuit design, engineering, and fabrication, see Warner (1965).

13. A semiconductor is an electric circuit component, such as a transistor (or chip) that is fabricated from a material that is neither a good conductor of electricity nor a good insulator. Pure silicon is a poor conductor of electricity, but by a process called "doping" a number of atoms of another substance can be introduced into the silicon crystal to alter the electrical properties. A transistor is made of three layers of silicon. Each layer is doped with impurities in such a way that electric current passing through the transistor can be influenced by the much smaller current applied to the middle layer. An *integrated circuit* is a single chip that has more than one active device on it, such as transistors, diodes, resistors, or capacitors, as part of an electric circuit. The first integrated circuits required that the connections between circuit elements be made by hand. There are three main types of integrated circuits: (1) memory chips, (2) microprocessors, and (3) microcomponents (Ruttan 2001, pp. 326–328).

14. Although the transistor is sometimes cited as an example of a "science push" invention, a clear demand-side incentive for the development of such a device was apparent at Bell Tele-

The relationship between the development of science and that of technology in the work of Shockley, Bardeen, and Brattain at Bell Laboratories was clearly much more complex than is implied by the linear model in which basic research precedes applied research, and applied research precedes technology development (figure 3.1). The Bell Laboratories group was involved simultaneously in advancing semiconductor theory and in semiconductor technology development. Although the work of Shockley, Bardeen, and Brattain was directed to the solution of an immensely important engineering problem, it was regarded as sufficiently fundamental that the team was awarded the Nobel Prize in Physics in 1956 (Shockley 1976).

Until the late 1950s transistors were discrete devices—each transistor had to be connected to other transistors on a circuit board. In the mid-1950s Texas Instruments, then the leader in silicon transistor production, initiated a research program under the direction of Jack Kilby to repackage the semiconductor products (transistors, resistors, and capacitors) as single components to reduce circuit interconnections. In 1958 Kilby's efforts resulted in a first very crude integrated circuit. The costs of assembling the separate components of Kilby's device by hand were too expensive for commercial application. At about the same time, however, Robert Noyce and Gordon Moore at Fairchild Semiconductor independently invented the integrated circuit. The invention, termed the *planar process*, involved incorporating very small transistors and resistors on a small sliver of silicon and adding microscopic wires to interconnected adjacent components.

The third major invention in the development of the semiconductor industry was the microprocessor. There are two types of integrated circuits that were critical in the development of computers. One, a memory chip, allows the computer to temporarily remember programs and other information. The other is the microprocessor, which processes the information rather than stores it.[15] The first micro-

phone Laboratories. The motives of Bell Laboratories for establishing the solid-state physics group were that major advances in the field were likely to be fruitful in improving communications technology. Shockley's own interests embraced both the prospect of advancing semiconductor theory and of developing a solid-state amplifier. The approach that Shockley and his associates at Bell undertook was to make an electronic amplifier of semiconductor material. This approach involved advancing the understanding of electron flow in semiconductor materials (Shockley 1976, pp. 618–619).

15. The critical invention that led to the development of the microprocessor was the metal oxide semiconductor (MOS). Initially the MOS transistor was much slower than the bipolar transistor but offered the offsetting benefit of simplicity, which enabled designers to put more transistors on an integrated circuit. The MOS technology made it possible to put the entire central processing unit on an integrated circuit. The first MOS transistor was built in 1960 by RCA under an Air Force contract. The transition from the bipolar point-contact technology to the metal oxide semiconductor technological trajectory occurred slowly. It took more than ten years to go from initial conception to commercial viability at Intel. For a detailed exposition of the history of MOS technology, see Bassett (2002).

processor was developed by Intel in 1969–1970. Technical progress in the integrated circuit era has moved along a trajectory toward increasing the density of circuit elements per chip: "In 1965 the cofounder of Intel, Gordon Moore, predicted that the number of transistors per integrated circuit would double every 18 months. This has come to be referred to as Moore's Law" (Jovanovic and Rousseau 2002; Flamm 2004, p. 170; see figure 5.1).

The potential military applications of semiconductors were immediately apparent. The transition between the initial invention of the transistor and the development of military and commercial applications of semiconductors and integrated circuits was guided and substantially funded by the Army Signal Corps. By 1953 the Army Signal Corps Engineering Laboratory was funding approximately 50 percent of transistor research and development at Bell Laboratories. The Signal Corps Engineering Laboratory at Fort Monmouth, New Jersey, developed the technology to replace hand soldering of components, a critical advance in the transition to mass production of transistor radios. In 1953 the Signal Corps underwrote the construction of a large Western Electric transistor plant at Lauredale, Pennsylvania. By the mid-1950s it had also subsidized facility construction by General Electric, Raytheon, RCA, and Sylvania. Funding was also provided for engineering development (Misa 1985, pp. 253–287).[16]

Demand created by the Minuteman II missile and the Apollo space projects (see chapter 7) pushed U.S. firms rapidly down the design and production learning curves (Alic et al. 1992, p. 257; Langlois and Steinmueller 1999, pp. 35–36). The diffusion of knowledge and the entry of new firms were encouraged by the military procurement policy of "second sourcing" to avoid becoming dependent on a single supplier: "By subsidizing engineering development and the construction of manufacturing facilities, and by leading the movement to standard operating characteristics, the military catalyzed the establishment of an industrial base" (Misa 1985, p. 28; see also Langlois and Steinmueller 1999, pp. 26–28). Demand for semiconductors continued to be dominated by direct procurement for military, nuclear power, and space applications as the need for increasingly powerful computers grew well into the 1970s (Wessner 2003, p. 13).

As already noted, progress in semiconductor development was beginning to open up the possibility of designing smaller and less expensive computers. When the PDP-8, the first computer to use integrated circuits, was introduced in 1965, it sold for $18,000 and could be rented for $525 per month. The transition from the minicomputer to the microcomputer—the *personal computer*—had to await

16. For a critical assessment of the role of military procurement in the development of semiconductors, see DeGrasse (1983, pp. 84–91).

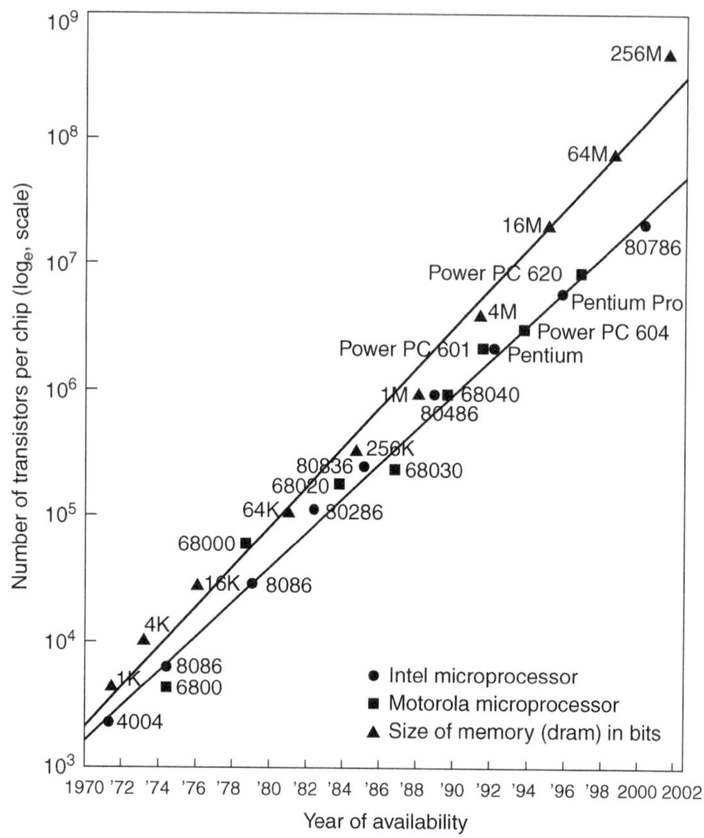

Figure 5.1. Transistor densities on microprocessors and memory chips. Adapted from G. Dan Hutchinson and Jerry D. Hutcheson, 1996, "Technology and Economics in the Semiconductor Industry," *Scientific American* 274 (January): 61. Used with permission of VLSI Research Inc.

Intel's 1969 development of the programmable chip. With the development and diffusion of the minicomputer and the microcomputer, the primary sources of demand for semiconductors, primarily memory chips and microprocessors, shifted strongly in the direction of commercial technology.[17]

17. It was the density of MOS circuitry that led to the dramatic reductions in the cost of semiconductor memory and microprocessor chips that made the personal computer possible. Because of their emphasis on speed, bipolar transistors continued to be the primary technology employed on the semiconductors by producers of mainframe computers, such as IBM, Control Data, and Cray, until well into the 1980s (Bassett 2002). I do not discuss in this chapter the development of the personal computer. I have reviewed the early history of personal computer development in Ruttan (2001, pp. 331–335).

The development of integrated circuits has had the effect of increasing the fixed costs of innovation, which became a barrier to the entry of new firms into the industry. In the 1970s U.S. producers shifted substantial production assembly capacity to low-wage developing countries, particularly Mexico, Taiwan, Singapore, Malaysia, and Korea. By the early 1980s Japanese semiconductor makers were turning out memory chips of much higher quality and for lower prices than even the leading U.S. producers, such as Intel. By the late 1980s Intel had withdrawn from the highly competitive memory chip sector to concentrate on the production of microprocessors (Ruttan 2001, pp. 328–331).

Concern by the defense agencies about the loss of U.S. competitive advantage led to the formation of SEMATECH, a consortium of semiconductor equipment manufacturing companies. A primary objective was to strengthen domestic innovation and capacity in the semiconductor equipment manufacturing industry. From its founding in 1987 until the mid-1990s, when it evolved from a U.S.-only consortium into a global consortium, SEMATECH received substantial funding from the Defense Advanced Research Projects Agency (DARPA).[18]

Supercomputers

In the early 1950s there was a relatively clear-cut distinction between computers designed primarily for business and those designed for scientific applications. The 1952 IBM 701 and the 1954 IBM 704 were regarded as scientific computers, whereas the 1953 IBM 702 and the IBM 705 were regarded primarily as business data processors.[19]

The AEC weapons laboratories were an early major source of demand for high-speed scientific computers. In the mid-1950s the weapons designers at Livermore National Laboratory estimated that "they would need a computer having one hundred times the power of any existing system" (MacKenzie 1996, p. 114). Bids were sought for such a machine from both IBM and Remington Rand. Livermore National Laboratory commissioned the development of LARC from Remington Rand, and the Los Alamos National Laboratory commissioned the Stretch from IBM.

18. I have discussed issues of international competition and the formation of SEMATECH and its troubled history in greater detail in Ruttan (2001, pp. 353–357). For a more recent perspective, see the several essays in Wessner (2003). Also, note that the name of the Advanced Research Projects Agency (ARPA) was changed to Defense Advanced Research Projects Agency (DARPA) in 1972 and renamed ARPA in 1993.

19. In this section I draw heavily on the work of MacKenzie (1996).

The LARC met the Livermore performance specifications, but not its expectations. Stretch did not meet even performance specifications. Only two LARC machines were ever built: the other went to the U.S. Navy's ship and reactor designers. Of the 8 Stretch machines that were sold, 4 were for nuclear research and development, 2 were for other military research and development, one went to the National Security Agency, and another one went to the U.S. Weather Bureau. Although both machines contributed to the development of supercomputer capacity, they were short-term financial disasters for the firms that built them (MacKenzie 1996, pp. 114–116).

As its work on the SAGE project was coming to completion, IBM faced several difficult problems. It was producing six different computer lines, all of which had incompatible operating systems. Competitors were beginning to make inroads into IBM's market share. Software was accounting for a greater proportion of the cost of computer systems. Many of these problems were resolved with the introduction of the IBM System 360 in 1965. The 360 family of computers used integrated circuits rather than transistors. They had large ferrite core memories with fast access times and multiprogramming that allowed many programs to run simultaneously, and an improved disk memory that allowed the machines to store more information in secondary memory than had previously been thought possible.

The 360 machines were designed to meet both business and scientific applications. No matter what size, all contained the same solid-state circuits and would respond to the same set of instructions. As it came on line, the System 360 platform became the industry standard for the rest of the 1960s and 1970s (Bresnahan 1999, pp. 227–228). The decision by IBM to commit to the 360 line had not been easy. It required an enormous technical and financial commitment: "IBM literally 'bet the company' on its 360 decision" (Katz and Phillips 1982, p. 218).

The alternative to the path followed by IBM was to design computers that would be substantially faster than the IBM 704 or any other IBM machine at floating point arithmetic.[20] The first machine that could properly be termed a *supercomputer* was the 1964 Control Data 6600.[21] It was designed by Seymour Cray, who would

20. A dominant objective in the design of supercomputers has been to achieve higher speed at floating point arithmetic. Speed, "conventionally expressed as the number of floating point operations ('flops') carried out per second, has increased from the thousands (kiloflops) in the 1950s to the millions (megaflops) in the 1960s to thousand million (gigaflops) in the 1980s, and may increase to the million millions (teraflops) by the end of the 1990s" (MacKenzie 1996, p. 100). For a more technical discussion, see MacKenzie (1996, pp. 166–175).

21. ERA was acquired by Remington Rand in 1952. In 1955 Remington Rand merged with the Sperry Corporation to form Sperry Rand. In 1957 William Norris, Seymour Cray, and several others left Sperry Rand and formed Control Data (MacKenzie 1996, p. 135). By the mid-

dominate supercomputer development for the next thirty years. Cray had designed Control Data's first computer, the CDC 1604, announced in October 1959. The 1604, built with transistors rather than vacuum tubes, was the world's fastest computer. Control Data had successfully challenged IBM at the technical level. It was unsuccessful, however, at challenging the commercial success of IBM—in terms of market share, revenues, or profits.

"Cray had no interest in business data processing," notes Donald MacKenzie, "and abhorred the complexity that arose from trying to cater to both scientific and business users" (1996, p. 136). The profitability of the 1604 enabled Cray to negotiate an arrangement with Control Data chairman William Norris to set up a laboratory in Chippewa Falls, Wisconsin, to conduct development work on the CDC 6600. As IBM chairman Thomas J. Watson, Jr., watched the 6600 dominate the high-speed computer market, "he asked, in an acerbic memo to his staff why Cray's team of 'only 34—including the night janitor'—had outperformed the computing industry's mightiest corporation" (MacKenzie 1996, p. 140).

Another Control Data team, working out of CDC headquarters in Arden Hills, Minnesota, worked on the development of the Control Data 3600, a highly successful series of computers, compatible with the 1604, with a primary orientation to the commercial market. By the early 1970s Control Data had grown into a large, diversified company. Its corporate plans included supercomputer development, but not at a pace that was satisfactory to Cray.

In 1972 Cray and several colleagues left Control Data to start up a new company, Cray Research, also located in Chippewa Falls. The customer base for the world's fastest computers was small—Cray estimated it at no more than fifty—primarily in the nuclear weapons laboratories, the National Aeronautics and Space Administration (NASA), the DOD, the National Center for Atmospheric Research, and the U.S. Weather Bureau: "The Cray-1 was as much a tour-de-force as the 6600" (MacKenzie 1996, p. 145). It was introduced in 1976. The first sale was to the Los Alamos nuclear weapons laboratory, and the second was to the National Center for Atmospheric Research in Boulder, Colorado. The Cray-1 was followed by a more advanced machine, the Cray-2. A less expensive and even faster machine using parallel processing, the X-MP, developed by staff members Lester Davis and Steve Chen, was introduced in 1982. By 1989 the X-MP, with almost 160 sales of different versions, was the Western world's most successful

1960s the computer industry consisted of IBM and the "seven dwarfs." The composition, in terms of share (percent) of sales, was as follows: IBM (65.3), Sperry Rand (12.1), Control Data (5.4), Honeywell (3.8), Burroughs (3.5), General Electric (3.4), RCA (2.9), and NCR (2.9). By the mid 1990s only IBM was still active in the computer industry under its 1965 corporate identity.

supercomputer. In 1990 Cray research announced a slower and significantly less expensive mini-supercomputer, the Y-MP. The success of the Cray-2 and of the X-MP and the Y-MP enabled Cray Research, now chaired by John Rollwagen, to edge its way into the ranks of one of the dozen leading computer manufacturers in the world.[22]

Seymour Cray's commitment to speed above all other objectives led to a second parting of ways. Resources were not available at Cray Research to complete the development of a Cray-3 and to pursue what were regarded as more promising development agendas. Rollwagen proposed the establishment of a Cray Research subsidiary, Cray Computer Corporation, to undertake the further development of the Cray-3. Cray moved his research team to Colorado Springs, Colorado; "In March 1995 the Cray Computer Corporation having failed to find a customer for the Cray-3 or a firm order for its successor the Cray-4, filed for bankruptcy protection" (MacKenzie 1996, p. 157). By 1995 neither Control Data nor Cray Research existed as independent firms.

At the time I began writing this chapter, the Japanese Earth Simulator, released in the spring of 2002 and designed specifically to support geoscience research and applications, was the world's most advanced supercomputer. Concern about the national security implications of loss of U.S. leadership led to a number of initiatives and studies, including an ARPA initiative to support the development of a new generation of economically viable high-productivity computer systems by 2007, and the formation of an NRC Study on the Future of Supercomputing. The NRC committee interim report identifies a number of nonmilitary sources of demand for more advanced supercomputing, such as bioinformatics, the large volume of data generated by the Human Genome Project, population genetics, and others. The committee notes that "in the United States, Japan and Europe, the majority of supercomputers have been purchased directly or indirectly using government funds, and the committee has no evidence that this is likely to change in the future" (National Research Council 2003, p. 28).[23]

22. By 1990 IBM had temporarily decided that high-end supercomputers represented a niche market in which it could no longer afford to compete. Control Data made the same decision in 1989 when it closed the ETA subsidiary that it had established to compete with Cray Research (MacKenzie 1996, p. 104).

23. Unlike computer technology development in the United States, that in Japan was driven to only a minor degree by military procurement. Beginning in 1960 the Japanese government designated computers as a "strategic industry" critical to the future economic development of the country. It took a quarter of a century of substantial government support, primarily by the Ministry of International Trade and Industry (MITI) for Japan to develop a computer industry fully competitive with the U.S. computer industry. The MITI objective in computer development from 1960 through the 1970s was to catch up to IBM. By the mid-1980s Japanese firms were

Software

A comprehensive treatment of the computer software industry would require a separate chapter. But, because of its intimate relationship with the development and diffusion of the computer, it is useful to briefly discuss the role of military procurement in software development.[24]

Before the 1960s computer software hardly existed as a distinct technology or industry. Early electronic computers, like their electromechanical business machine precursors, were programmed by rewiring: "Software was effectively born with the development by von Neumann of his conceptual architecture for computers. . . . But even after the von Neumann scheme became dominant . . . software remained closely bound to hardware. During the 1950s, the organization designing the hardware generally designed the software as well" (Langlois and Mowery 1996, pp. 55–56). During this period large military procurement contracts such as the SAGE air defense system played a particularly important role in embodied software development.[25] Mowery indicates that even as late as the early 1980s the DOD accounted for the largest share of the U.S. software market. Military procurement of contract software accounted for approximately half of the U.S. traded software market (Mowery 1999, p. 145). Because of the specialized function of such development efforts, however, there was often limited direct spillover to commercial application. Langlois and Mowery argue, for example, that one of the greatest contributions of SAGE was the training of a large cadre of skilled systems programmers (Langlois and Mowery 1996, p. 59).[26]

---

introducing a generation of supercomputers that approached, and in some cases exceeded, the performance of leading U.S. producers of supercomputers used for scientific purposes (Ruttan 2001, pp. 344–347). In late 2004 IBM announced that it had designed and was building the world's fastest computer (Crissey 2004).

24. In this section I draw primarily on Mowery (1996); Steinmueller (1996, pp. 15–52); Langlois and Mowery (1996, pp. 53–85); and Mowery and Rosenberg (1998, pp. 1953–1966); Mowery (1999); and Mahoney (2002). For a highly personal account of the stages of software development, see Glass (1998, pp. 12–26).

25. The defining moment for the software contracting industry came in 1956 when the RAND Corporation established the Systems Development Corporation (SDC) to assume responsibility for computer program development for the SAGE air defense project. When the SAGE project was winding down in 1960, SDS continued to grow by taking on other military and defense-related projects. By 1963 it had an annual income of $57 million generated by contracts with the Air Force, NASA, the Office of Civil Defense, ARPA, and other defense and defense-related-sector projects (Campbell-Kelley 2003, p. 41).

26. At a time when the entire population of programmers in the United States was about 1,200, some 700 were employed on the SAGE project. "The first and classic civilian real-time project was the IBM–American Airlines SABRE airline reservation system. Though SABRE did not become fully operational until 1964, it was the outcome of more than 10 years of planning,

In the middle and late 1960s, three events contributed to the "disintegration" of the computer and software industries (Steinmueller 1996, pp. 24–26). One was the introduction beginning in 1964 of the IBM System 360 family of computers. The System 360 gave independent software service companies and vendors an opportunity to develop and market the same product to a variety of users. The second event was the decision by IBM in 1969 to unbundle the sale of hardware and software. This provided an incentive for independent software developers to produce software compatible with IBM products. The third important event was the development of the minicomputer industry. Minicomputers made it possible for small organizations to purchase and operate their own computers. Each of the many different uses of minicomputers—"as primary computers in smaller organizations, as 'front ends' for mainframes, in data communication systems, and in process central systems—required very different software" (Steinmueller 1996, p. 28). Since the early 1980s the success of the personal computer has led to an explosive growth in the mass-market software products industry.[27]

Defense-related research support for the computer software industry has differed substantially from the patterns followed in the computer and semiconductor industries. In the case of computers and semiconductors, defense agency research support occurred primarily through direct procurement. "In software, by contrast, defense-related R&D funded computer science in much of the 1950s and 1960s . . . was directed to facilitating advances in fundamental knowledge of computer architecture, software languages, and design that found application in both the civilian and defense sectors. . . . Military-civilian spillovers in software occurred as a result of defense-related R&D spending rather than from direct software procurement" (Langlois and Mowery 1996, p. 14; see also Flamm 1987, pp. 42–92; Norberg and O'Neill 1996).[28]

In the middle and late 1960s, computer science departments were established at Stanford, Carnegie Mellon, and MIT. From the early 1970s through the mid-

technical assessment, and system building" (Campbell-Kelley 2003, p. 42). SABRE was a direct spin-off of the SAGE project (National Research Council 1999, p. 94).

27. A number of the important software innovations that contributed to the rapid adoption of the personal computer beginning in the early 1980s were spin-offs from the MIT Project MAC (later Laboratory for Computer Science) funded by the ARPA Information Processing Office beginning in 1963. Project MAC's development of software to compose and edit programs and documents online laid the groundwork for word processors. Other Project MAC software spin-offs included the spreadsheet, early versions of Internet protocols for the personal computer, and the UNIX operating system (National Research Council 1999, pp. 103–105).

28. "Between 1976 and 1995 DOD provided some 60 percent of total federal research funding for computer science and over 75 percent of total funding in electrical engineering" (National Research Council 1999, p. 56). For a listing of defense-to-civilian "spillovers" in the U.S. software industry between 1950 and 1975, see Flamm (1988, pp. 266–268).

1980s, more than half of academic computer science research and development (R&D) was accounted for by defense-related agencies. After the mid-1980s defense-related support for computer science at academic institutions shifted toward application, and the defense share of federal support for computer science funding declined from almost 60 percent to less than 30 percent (Aspray and Williams 1994; Langlois and Mowery 1996, p. 71).[29]

The history of technology development in software production has differed from that in computer production in one other important respect. Technical change in computer production and performance has led to very substantial growth in labor productivity in the production of computers (Jorgenson 2001; Ruttan 2001, pp. 357–362). In contrast, software production has remained exceedingly labor intensive. In 1955 software constituted less than 20 percent of the costs of a computer system. By the mid-1980s software costs were estimated at upward of 90 percent of the cost of installing a system (Boehm 1973; Mahoney 2002). By the late 1990s slow productivity growth in software design and production had become an obstacle to productivity growth in the computer industry and in the computer-intensive defense and commercial sectors.

Perspective

The invention and early development of the electronic digital computer were supported entirely by Army and Navy contracts. Defense and defense-related agencies were the dominant supporters of the scientific and technical innovations and the primary market for computers until at least the early 1960s. These agencies, particularly the laboratories of the AEC (later the National Energy Department), greatly influenced the direction of technology development and have been the primary source of demand for increasingly powerful supercomputers (MacKenzie 1996, p. 117).

In the case of semiconductors, the DOD went beyond procurement to support research and development at Bell Laboratories and to subsidize the private-sector facilities development to create an industrial base and to assure a competitive structure in the semiconductor industry. In the case of software and artificial intelligence, the military supported fundamental research and invested in the establishment of academic computer science training.

29. The initial development of computers preceded the development of computer science. For a brief account of the dialectical relationship between the development of computers and computer science, National Research Council (1999, pp. 184–197).

MacKenzie, concluding his article on nuclear weapons laboratories and supercomputing, asks the following question: What does it mean for an institution to have influenced the development of an area of technology? His answer is that "without Los Alamos and Livermore we would doubtless have a category of supercomputing—a class of high performance computers—but the criterion of performance that would have evolved would have been much less clear cut" (1996, p. 126).

Others have been less cautious. Kenneth Flamm has argued that, even without the impetus of military procurement, the modern electronic computer would probably have been developed, but "the pace of development would have been far slower" (1988, p. 251). Paul Edwards insists that, without the very large funding and the sense of urgency associated with the war effort, computer development in the United States would have been delayed for at least a decade (1996, p. 52).

My own reading of the literature is that the development and commercialization of computers—mainframes, minicomputers, and microcomputers, would have occurred even more slowly than suggested by Flamm and Edwards. Without the impetus of the SAGE project, for example, a cautious IBM and a financially constrained Remington Rand would have substantially delayed the investment necessary for the emergence of the technology that enabled the development of mainframe computers. Efforts to develop increasingly powerful computers played an important role in shaping the architecture and performance of mainstream computers ranging from workstations to personal computers.

It has been difficult until quite recently to assess either the social rates of return to investment in the development of computers, or the contribution of computers to U.S. economic growth (Flamm 1987, pp. 223–239; Jorgenson and Stiroh 1999; Ruttan 2001, pp. 357–362). There can be little question, however, that adoption of business and personal computers contributed importantly to the recovery of productivity growth in the U.S. economy, beginning in the early 1990s. And there can be little doubt that, in the absence of the impetus for development and commercialization associated with military procurement, significant contributions of computer and related information technologies to the growth of the U.S. economy would have been delayed until well into the twenty-first century.

ACKNOWLEDGMENTS

I am indebted to Arthur Norberg, Robert Seidel, and Jeffrey Yost for comments on an earlier version of this chapter.

REFERENCES

Alic, J. A., L. M. Branscomb, H. Brooks, A. B. Carter, and G. L. Epstein. 1992. *Beyond Spinoff: Military and Civilian Technology in a Changing World.* Boston: Harvard Business School Press.

Aspray, W., and B. O. Williams. 1994. "Arming American Scientists: NSF and the Provision of Scientific Computing Facilities for Universities, 1950–1973." *IEEE Annals of the History of Computing* 16: 60–74.

Bashe, C. J., L. R. Johnson, J. H. Palmer, and E. W. Pugh. 1986. *IBM's Early Computers.* Cambridge, MA: MIT Press.

Bassett, R. K. 2002. *To the Digital Age: Research Labs, Start-Up Companies, and the Rise of MOS Technology.* Baltimore: Johns Hopkins University Press.

Boehm, B. 1973. "Software and Its Impact: A Quantitative Assessment." *Datamation* (May): 48–59.

Bresnahan, T. F. 1999. "Computing." In *U.S. Industry in 2000: Studies in Comparative Performance,* ed. D. C. Mowery. Washington, DC: National Academy Press, 215–244.

Campbell-Kelley, M. 2003. *From Airline Reservations to Sonic the Hedgehog: A History of the Software Industry.* Cambridge, MA: MIT Press.

Campbell-Kelley, M., and W. Aspray. 1996. *Computer: A History of the Information Machine.* New York: Basic Books.

Ceruzzi, P. E. 2003. *A History of the Modern Computer,* 2nd ed. Cambridge, MA: MIT Press.

Cortada, J. W. 1993. *Before the Computer: IBM, NCR, Burroughs and Remington Rand and the Industry They Created, 1865–1956.* Princeton, NJ: Princeton University Press.

Crissey, M. 2004. "It's Official: IBM Fastest in World." *Minneapolis Star Tribune,* 9 November, D2.

DeGrasse, R. W., Jr. 1983. *Military Expansion, Economic Decline: The Impact of Military Spending.* Armonk, NY: Sharpe.

Edwards, P. N. 1996. *The Closed World: Computers and the Politics of Discourse in Cold War America.* Cambridge, MA: MIT Press.

Flamm, K. 1987. *Targeting the Computer: Government Support and International Competition.* Washington, DC: Brookings Institution Press.

———. 1988. *Creating the Computer: Government, Industry and High Technology.* Washington, DC: Brookings Institution Press.

———. 2004. "Moore's Law and the Economics of Semiconductors." In *Productivity and Cyclicality in Semiconductors: Trends, Implications, and Questions,* ed. D. W. Jorgenson and C. W. Wessner. Washington, DC: National Academies Press, 150–170.

Glass, R. L. 1998. "Software Reflections—A Pioneer's View of the History of the Field." In *In the Beginning: Personal Recollections of Software Pioneers,* ed. R. L. Glass. Los Alamitos, CA: IEEE Computer Society.

Hughes, T. P. 1998. *Rescuing Prometheus.* New York: Random House.

Hutchinson, G. D., and J. D. Hutcheson. 1996. "Technology and Economics in the Semiconductor Industry." *Scientific American* 274 (January): 61.

Jorgenson, D. W. 2001. "Information Technology and the U.S. Economy." *American Economic Review* 91: 1–32.

Jorgenson, D. W., and K. J. Stiroh. 1999. "Information Technology and Growth." *American Economic Review* 89: 109–115.

Jorgenson, M. R. 1996. "Monopoly and Markets in the U.S. Computer Industry to 1970: IBM and U.S. Government Technology and Antitrust Policy." Minneapolis: University of Minnesota, Department of Sociology. Mimeograph.

Jovanovic, B., and P. J. Rousseau. 2002. "Moore's Law and Learning by Doing." Cambridge, MA: National Bureau of Economic Research Working Paper 8762.

Katz, B. G., and A. Phillips, 1982. "The Computer Industry." In *Government and Technical Progress: A Cross-Industry Analysis*, ed. R. R. Nelson. New York: Pergamon Press, 162–232.

Langlois, R. N. 1992. "External Economies and Economic Progress: The Case of the Microcomputer Industry." *Business History Review* (Spring): 1–50.

Langlois, R. N., and D. C. Mowery. 1996. "The Federal Government's Role in the Development of the U.S. Software Industry." In *The International Computer Software Industry: A Comparative Study of Industry Evolution and Structure*, ed. D. C. Mowery. New York: Oxford University Press, 53–85.

Langlois, R. N., and W. E. Steinmueller. 1999. "The Evolution of Competitive Advantage in the Worldwide Semiconductor Industry, 1947–1996." In *Sources of Industrial Leadership: Studies of Seven Industries*, ed. D. C. Mowery and R. R. Nelson. Cambridge: Cambridge University Press, 19–78.

Lee, J.A.N. 2002. "Pioneers in Computing." In *From 0 to 1: An Authoritative History of Modern Computing*, ed. A. Akera and F. Nebeker. New York: Oxford University Press, 76–87.

Levine, R. C. 1982. "The Semiconductor Industry." In *Government and Technical Progress: A Cross-Industry Analysis*, ed. R. R. Nelson. New York: Pergamon Press, 7–100.

MacKenzie, D. 1996. *Knowing Machines: Essays on Technical Change*. Cambridge, MA: MIT Press.

Mahoney, M. S. 2002. "Software: The Self-Programming Machine." In *From 0 to 1: An Authoritative History of Modern Computing*, ed. A. Akera and F. Nebeker. New York: Oxford University Press, 91–100.

Misa, T. J. 1985. "Military Needs, Commercial Realities and the Development of the Transistor, 1948–1958." In *Military Enterprise and Technological Change: Perspectives on the American Experience*, ed. M. R. Smith. Cambridge, MA: MIT Press, 253–298.

Mowery, D. C. 1983. "Innovation, Market Structure, and Government Policy in the American Semiconductor Electronic Industry." *Research Policy* 12: 183–197.

———, ed. 1996. *The International Computer Software Industry: A Comparative Study of Industry Evolution and Structure*. New York: Oxford University Press.

———. 1999. "The Computer Software Industry." In *Sources of Industrial Leadership:*

Studies of Seven Industries, ed. D. C. Mowery and R. R. Nelson. Cambridge: Cambridge University Press, 133–168.

Mowery, D. C., and N. Rosenberg. 1998. *Paths of Innovation: Technological Change in 20th Century America.* Cambridge: Cambridge University Press.

National Research Council. 1999. *Funding a Revolution: Government Support for Computing Research.* Washington, DC: National Academy Press.

———. 2003. *Interim Report on the Future of Supercomputing.* Washington, DC: National Academy Press.

Nelson, R. R. 1962. "The Transistor." In *The Rate and Direction of Economic Activity: Economic and Social Factors,* ed. R. R. Nelson. Princeton, NJ: Princeton University Press, 549–583.

Norberg, A. L. 1993. "New Engineering Companies and the Evolution of the United States Computer Industry." *Business and Economic History* 22: 181–193.

———. 2002. "The Modern Computing Business." In *From 0 to 1: An Authoritative History of Modern Computing,* ed. A. Akera and F. Nebeker. New York: Oxford University Press, 163–176.

Norberg, A. L., and J. E. O'Neill. 1996. *Transforming Computer Technology: Information Processing for the Pentagon, 1962–1986.* Baltimore: Johns Hopkins University Press.

Pugh, E. W. 1984. *Memories That Shaped an Industry: Decisions That Lead to IBM System 1360.* Cambridge, MA: MIT Press.

Redmond, K. C., and T. M. Smith. 2000. *From Whirlwind to MITRE: The R&D Story of the SAGE Air Defense Computer.* Cambridge, MA: MIT Press.

Rifkin, G., and G. Harrar. 1988. *The Ultimate Entrepreneur: The Story of Ken Olson and Digital Equipment Corporation.* Chicago: Contemporary Books.

Riordan, M., and L. Hoddeson. 1997. *Crystal Fire: The Birth of the Information Age.* New York: Norton.

Ruttan, V. W. 2001. *Technology, Growth, and Development: An Induced Innovation Perspective.* New York: Oxford University Press.

Seidel, R. W. 2002. "Government and the Emerging Computer Industry." In *From 0 to 1: An Authoritative History of Modern Computing,* ed. A. Akera and F. Nebeker. New York: Oxford University Press, 189–201.

Shockley, W. 1976. "The Path to the Construction of the Junction Transistor." *IEEE Transactions on Electronic Devices* ED-23 (July): 597–620.

Shurkin, J. 1984. *Engines of the Mind: A History of the Computer.* New York: Norton.

Slater, R. 1987. *Portraits in Silicon.* Cambridge, MA: MIT Press.

Steinmueller, W. E. 1996. "The U.S. Software Industry: An Analysis and Interpretive History." In *The International Computer Software Industry: A Comparative Study of Industry Evolution and Structure,* ed. D. C. Mowery. New York: Oxford University Press, 15–52.

Teal, G. K. 1976. "Single Crystals of Germanium and Silicon—Basic to the Transistor and Integrated Circuit." *IEEE Transactions on Electronic Devices* ED-23 (July): 621–639.

Tomash, E., and A. A. Cohen. 1979. "The Birth of ERA; Engineering Research Associates, Inc., 1946–1955." *Annals of the History of Computing* 1 (October): 83–97.

Usselman, S. W. 1993. "IBM and Its Imitators: Organizational Capabilities and the Emergence of the International Computer Industry." *Business and Economic History* 22: 1–35.

Warner, R. M., ed. 1965. *Integrated Circuits: Design Principles and Fabrication.* New York: McGraw-Hill.

Watson, T. J., Jr., with P. Petre. 1990. *Father, Son & Co.: My Life at IBM and Beyond.* New York: Bantam Books.

Wessner, C. W., ed. 2003. *Securing the Future: Regional and National Programs to Support the Semiconductor Industry.* Washington, DC: National Academies Press.

6

Inventing the Internet

The development of the Internet involved the transformation of a computer network that had initially been established in the late 1960s by the Advanced Research Projects Agency (ARPA) of the U.S. Department of Defense (DOD).[1] The decision to support development of ARPANET followed several earlier successful DOD efforts in the field of computer communication. The Whirlwind computer developed in the early 1950s for the Semi-Automatic Ground Environment air defense system enabled operators to interact with the data being processed and display information on a computer screen (chapter 5). The DOD was interested in the development of a more sophisticated system that could ensure survival of the communications system after an attack that might disable any single control station.

The decision also reflected the personal interest of Joseph Licklider, the first director of the ARPA Information Processing Techniques Office (IPTO), in man-machine interaction.[2] Licklider was impatient with the "batch process" used to process data on existing mainframe computers. To solve a problem with a computer, the researcher had to first formulate the problem and then turn to a professional programmer to program the problem for the computer. After the program was written in computer language, it was submitted to the operator of a centrally housed computer, who placed the program in a queue. The computer then proc-

1. I draw more heavily on Abbate (1999) than I can possibly acknowledge. I also draw heavily on Norberg and O'Neill (1996) and Hughes (1998). As mentioned in chapter 5, the name of the Advanced Research Projects Agency (ARPA) was changed to Defense Advanced Research Projects Agency (DARPA) in 1972 and renamed ARPA in 1993. In this chapter, I have followed the practice of using ARPA rather than DARPA.

2. ARPA was established in 1958 to provide oversight of research and development in the DOD and to sponsor frontier technology development with potential application to all three services—Army, Navy, and Air Force. For the early history of ARPA and its IPTO, see Norberg and O'Neill (1996, pp. 5–18, 25–53, 153–155).

essed the information and printed out the results (Hughes 1998, p. 261): "Batch processing rationalized the flow of input to the computer, but it was frustrating and inefficient for the programmer" (Abbate 1999, p. 24).[3] Licklider was concerned that computer technology serve the needs of the user rather than force the user to adapt to the machine (Licklider 1960).[4]

In this chapter I trace the development of the Internet, from the initial interest in developing a technology "that would enable big computers to talk to each other," through the invention and privatization of the Internet. I also give particular attention to the institutional innovations that were involved in both the invention and the adoption of the Internet.[5]

Creating ARPANET

Licklider initially visualized a system of "time-sharing" in which a single computer located at a central location would be accessed by a number of users with individual terminals connected to the central computer by telephone (Norberg and O'Neill 1996, pp. 88–94). Long-distance time-sharing with users connected to central terminals using dial-up telephone links would economize on the use of the central terminals—the scarce resource in the system. Analogies were drawn to the central power station in an electrical grid. No one anticipated the central-computer constraint would be eliminated by the development of mini- and microcomputers (chapter 5).

In 1966 Robert Taylor, then head of IPTO, secured the services of Lawrence

3. It was also time-consuming and frustrating to the user. I recall, when I was a visiting assistant professor at the University of California at Berkeley in 1958–1959, tabulating county data from the agricultural census, converting the data to logarithms by hand, delivering the data sheets to a young man in a white coat at the computer center, and waiting for more than a week for a printout of the regression results.

4. I do not address the ARPA research program in artificial intelligence (AI) in this chapter. The rationale for support of research in the area of AI, as articulated by Licklider, was that AI was central to the ARPA mission because it was a key to the development of advanced command-and-control systems. In 1983 ARPA initiated an even more ambitious AI program under the rubric of Strategic Control. The National Research Council credits ARPA support for AI research with important technical accomplishments. ARPA and NSF support for AI research is also credited with contributing importantly to the development of the discipline of computer science. In retrospect, however, it is hard to avoid a conclusion that the objectives of AI articulated by its early proponents and supporters were wildly optimistic (Norberg and O'Neill 1992, pp. 297–351; National Research Council 1999, pp. 198–225; Roland 2002).

5. I have elaborated elsewhere a model in which institutional innovations are induced by changes in resource endowments and technology, and in which technical change is induced by changes in resource endowments and institutions (Ruttan 2003, pp. 3–30).

Roberts, a Massachusetts Institute of Technology (MIT) Lincoln Laboratory researcher who had already connected a Lincoln Laboratory computer to one at the RAND Corporation in Santa Monica. Roberts was given a mandate to build a large multicomputer network that would interconnect the time-sharing computers at seventeen academic, industrial, and government computer centers funded by ARPA.

Packet Switching

Taylor had initially been committed to using relatively low-speed telephone lines to connect the computers at the several centers. At a computing symposium held at Gatlinburg, Tennessee, in October 1967, he was made aware by Roger Scantlebury of the British National Physical Laboratory (NPL) of the very substantial progress that had been made in a technique of message transmission termed *packet switching.*

A proposal for the development of a distributed communication system with fast end-to-end transmission of messages and small inexpensive switches was initially advanced by Paul Baran, a young engineer working at RAND, in a series of papers written in the late 1950s and early 1960s (Baran 1964). Messages would be broken into small "packets" and routed over the distributed system automatically rather than manually. He envisaged an all-digital network with computerized switches and digital transmission. Packet switching appealed to Baran because it seemed to meet the requirement of a survivable military communication system better than the more highly centralized system being built for the DOD by Bell Telephone Laboratories. The proposal became embroiled in a conflict with the new Defense Communication Agency (DCA) and was withdrawn by its Air Force sponsors.[6] It was only the commitment of very large financial and technical resources that were available to ARPA that assured the success of the packet switching technology.

The computer hardware and operating systems available in the mid-1960s tended to be relatively specialized. Contractors who needed access to different modes of computing had to travel to another site or acquire multiple machines. IPTO was continually under pressure from the centers to support the purchase of

6. An alternative packet switching system with commercial rather than military objectives in mind was developed under the direction of Donald Davies of the British NPL, independent of the work by Baran, in the mid-1960s. Bureaucratic and resource constraints prevented the realization of either Baran's or Davies's objectives. For a more detailed account of the convoluted history of the development of packet switching in the United States and the United Kingdom, see Abbate (1999, pp. 7–41).

additional computers. For Taylor and Roberts, a network system appeared to be both an opportunity to advance research in a new branch of computer science and to save ARPA money on computer facilities. At a planning session at the University of Michigan in 1967, Roberts laid out his vision of a system in which the host computers would be interconnected by small interface computers, thus enabling host computers with different characteristics to be able "to speak to each other." The proposal was initially resisted by several of the university-based principal investigators who were committed to the development of their own software. But because all the centers were dependent on ARPA support, Roberts was able to insist that all twelve sites link their computers to the network (figures 6.1 and 6.2)—"whether they wished to or not" (Abbate 1999, p. 46).

ARPA proceeded to award a contract for the development of a computer—an interface message processor (IMP)—that could route message packets along alternative routes to Bolt, Beranek and Newman (BBN), a small high-technology firm located near Cambridge, Massachusetts. Both the development of the software that would route the packets through alternative connections and the engineering design problems turned out to be more difficult than was anticipated. But the BBN IMP team was successful in developing the basic elements of the Internet nine months after the contract had been let.[7]

In his assessment of the accomplishments of the "IMP boys," Thomas Hughes insists that "future historians, fully aware of the remarkable development of the worldwide Internet, following hard upon the path-breaking ARPANET may someday compare the inventive success of the small BBN group to the achievement of Thomas Edison and his small band of associates who invented the electric lighting system" (1998, p. 278). As in the case of Edison's research at Menlo Park, there was intense dialectical interaction between advances in science and advances in technology. Sometimes invention was informed by science, and at other times invention came first, followed by scientific insight (figure 3.1).

By the end of 1971, the network consisted of fifteen sites, including sites run by the Air Force and the National Bureau of Standards. Yet the system was still not fully operational. Because of the large effort involved, host system operators were slow to build the special-purpose hardware interface between their computer and its IMP. Roberts and his associates decided that a dramatic gesture would be required to "galvanize the network community into making a final push to get their resources online." Abbate notes, "They arranged to demonstrate ARPANET's ca-

7. For more complete accounts, see Norberg and O'Neill (1996, pp. 162–172), Hafner and Lyon (1996), and Hughes (1998, pp. 275–282).

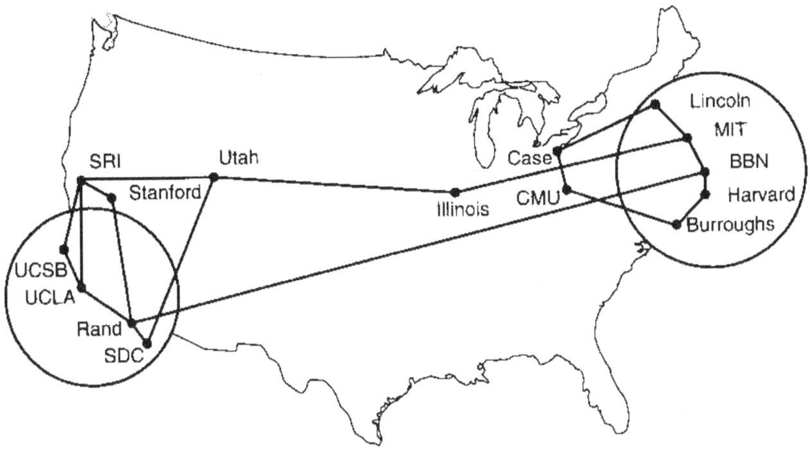

Figure 6.1. The main Information Processing Techniques Office (IPTO) Research Centers at the time of ARPANET's creation. *Source:* Janet Abbate, 1994, *Inventing the Internet* (Cambridge, MA: MIT Press), 45. Copyright © 1994. Reprinted with permission of The MIT Press.

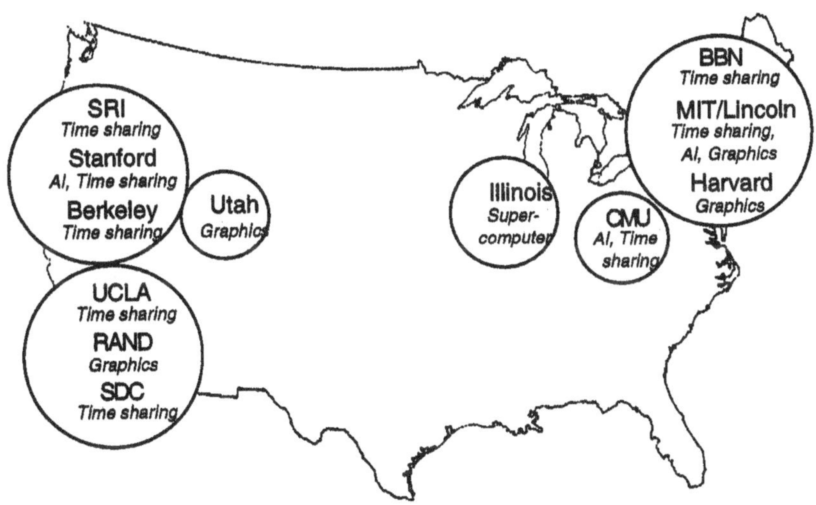

Figure 6.2. Map of the fifteen-node ARPANET in 1971. *Source:* Janet Abbate, 1994, *Inventing the Internet* (Cambridge, MA: MIT Press), 45. Copyright © 1994. Reprinted with permission of The MIT Press.

pabilities at the First International Conference on Computer Communication, which was to be held that October in Washington" (1999, p. 78).

The demonstration of the Internet at the 1972 International Conference on Computer Communication was the defining moment in the diffusion of use of the Internet. It finally convinced the skeptics in the computer and telephone industries that packet switching could become a viable commercial technology. It could no longer be considered simply a potential defense application or a research tool. Although its potential capacity as a communication tool was apparent, at least to those who had participated in its development, neither the DOD sponsors of the research nor the members of the design team anticipated that its primary use would be for personal and commercial e-mail rather than for transmitting data or for research collaboration. Nor was it apparent that it would take an additional quarter century to resolve the technical and institutional problems necessary to release the potential of the Internet.

Learning by Using

Accessing and using the Internet in the early 1970s posed a series of challenges. In order for a site to get an ARPANET connection, someone had to have a research contract with ARPA or another government agency approved by ARPA. Adding a new site was constrained by the high cost of establishing access, by the need to be affiliated with an ARPA research group, and by access to skilled programmers capable of creating and maintaining the host software.[8]

Once a university or company was connected to ARPANET, however, access controls were quite loose. Almost anyone who had access to an account on the connected computer could use network applications. But once a user had access to ARPANET, additional problems were encountered. Search tools and address books did not exist. Arrangements to get access to remote computers had to be arranged. Many of the host computers had their own command languages, data formats, and specialized hardware. Compatibility problems turned out to be much

8. A network member not funded by ARPA had to pay the cost of setting up a new node, "estimated in 1972 to be somewhere between $55,000 and $107,000": "Once a site was approved, ARPA had to order a new IMP from Bolt, Beranek and Newman, direct the Network Analysis Corporation to reconfigure the network to include the new node, and arrange for AT&T for a telephone link between the new node and the rest of ARPANET. The new host would be responsible for providing hardware and software for the host-IMP interface and for implementing the host protocol, NCP on its computer(s)—a task that might represent a year's work for a programmer" (Abbate 1999, pp. 84–85).

more difficult to resolve than the IMP boys or anyone else working on the system had anticipated.

During the 1970s important changes in hardware, software, configuration, and applications were initiated by ARPANET in response to users' concerns.[9] The decentralized ARPA environment created an opportunity for users to both vent their frustrations and devise new applications. Abbate discusses several areas where users' experiments, often using ARPA funding, made significant contributions. One was the development of terminal interface systems. Pressure from University of California, Los Angeles, which wanted to connect two computers to their IMP, led Roberts to authorize BBN to modify the IMP to handle more than one computer, and later to develop a way of connecting sites without host computers to connect directly to ARPANET (termed the terminal IMP or TIP). A second was the development of new communication paths to facilitate connection with local area networks (LANs). ARPANET was designed to connect distant computers. When users began sending data between computers at the same site, the practice led, to the surprise of ARPA and BBN staff, to the establishment of LANs.

Another important user-led development supported by ARPA was the effort at the University of Hawaii to develop radio packet switching terminals as an alternative to using leased telephone lines to link computers at the several university campuses. The principles developed in this project were later employed by Robert Metcalf, then working at the Xerox Palo Alto Research Center, to develop a random access network that used cable rather than a radio channel as the transmission medium. This system, initially termed *Aloha Alto*, later dubbed *Ethernet*, became the standard technique for LAN systems.[10]

Shortly after the 1972 demonstration of ARPANET at the International Conference on Computer Communications, Roberts left ARPA to head Telnet, BBN's commercial spin-off from ARPANET. At about the same time, Robert Kahn, who had organized the ARPANET demonstration, was appointed to the position of program manager at ARPA's IPTO. Kahn initiated a second Aloha spin-off project to build a local packet radio network (PRNET) linking ARPA contractors in the San Francisco Bay area. The objective was to test the (PRNET) technology to enhance the communication capacity of the military command-and-control system.

9. See Abbate (1999, pp. 83–111), and Norberg and O'Neill (1996).

10. ARPA management was not always responsive to users' concerns. Abbate (1999, pp. 90–96) notes that ARPA managers opposed the development of upper-level protocols because they were concerned that the network might slip out of their control.

A second Khan initiative was the development of packet switching for transmission of data via INTELSAT I (see chapter 7). The objective of the satellite packet switching network (SATNET) was to support both network research and the transmission of seismic monitoring data from Soviet nuclear weapons testing (Norberg and O'Neill 1996, pp. 180–182).

An important objective in the development of ARPANET had been to facilitate resource sharing. By the mid-1970s it was becoming apparent that running programs at remote sites was being done much less intensively than had been initially anticipated. The ideal of distributed computing and resource sharing made economic sense only when most ARPA researchers were using mainframe computers: "Had ARPANET's only value been as a tool for resource sharing the network might be remembered today as a minor failure rather than a spectacular success. But the network's users unexpectedly came up with a new focus for network activity: electronic mail" (Abbate 1999, p. 106).

By the mid-1970s ARPA was operating three separate packet switching networks: ARPANET, PRNET, and SATNET. Operational branches of the U.S. military and a number of European institutions were linked via ARPANET. The British Post Office and the Norwegian Telecommunications Authority linked sites in England and Norway to SATNET sites in the United States. Packet size and transmission speeds differed among the several networks. "No one in the field of computing had ever attempted to connect such dissimilar systems, and there were no models from which to work. As Kahn began thinking about ways to address the general problem of interconnecting heterogeneous networks, he set in motion what would become the Internet program" (Abbate 1999, p. 122). It was no longer enough for network builders to design a system that would facilitate communication among a set of computers; they now had to consider how different networks could interact.

Designing the Internet

In early 1973 Kahn explored with Vinton Cerf of Stanford University, one of the original designers of the ARPANET host protocol, the problems associated with the design of a system that would link ARPA's various networks. Cerf and Kahn then collaborated on a paper outlining the basic architecture for an ARPA Internet. In the paper they identified two basic questions: "First, if the packet radio network were to provide reliable connections with the host computers, it would need a host protocol that could compensate for its error-prone transmission medium. What would that host protocol look like? Second, what kind of mechanism could provide an interface between two distinct networks such as PRNET and ARPA-

NET?" (Cerf and Kahn 1974; Abbate 1999, p. 122). Kahn first contracted with Cerf to work out detailed specifications of the system, and then in 1976 he convinced him to come to ARPA as program manager for network projects.

Kahn and Cerf conferenced and consulted widely with computer researchers and with staff of other national network projects (the British NPL and French Cyclades networks) during the mid-1970s. Supporters of the Cyclades and Ethernet systems were concerned that the ARPANET Network Control Program (NCP) did not contain an error recovery mechanism. It was decided that the alternative Transmission Control Protocol (TCP), on which Cerf and Metcalf had earlier collaborated, would be employed to provide an orderly, error-free flow of data from host to host.

As an answer to the question of how to connect the networks physically, Cerf and Kahn proposed the creation of special host computers called "gateways": "A gateway would be connected to two or more networks and . . . all inter-network traffic would flow through these gateways." The network designers also "had to devise a system of host addresses that would enable packets to be directed to a particular host on a particular network." A hierarchical address scheme—the system we now use on our personal computers—facilitated the division of labor between gateways and local networks (Abbate 1999, p. 129).

Although the system worked out by Cerf, Kahn, and their colleagues met the project's original specifications, the system was criticized for imposing redundant functions on the gateways. It was not until 1977 that ARPA was able to demonstrate its first multinetwork connection. Packets were sent "from a van on a California freeway through PRNET to an ARPANET gateway, then through the ARPANET to a SATNET gateway on the East Coast, over SATNET to Europe, and finally through the ARPANET to California. . . . The successful three-way interconnection of the ARPANET, PRNET, and SATNET represented the beginning of the Internet as an operational system" (Abbate 1999, p. 131, figure 7.3).

Institutional Innovation

As early as 1972, following the demonstration of the technical feasibility of an ARPANET, ARPA began exploring the possibility of transferring ARPANET management to a commercial carrier or another government agency. Efforts to interest American Telephone and Telegraph (AT&T) were unsuccessful. In 1975 a decision was made to transfer operational responsibility to the DCA. ARPA would continue to provide funding and technical direction. Access would be restricted to DOD users and government contractors approved by the DCA.

The DCA immediately began to reorient the network away from its research focus toward military operations. This helped to overcome the DCA "not invented here" syndrome and resulted in more rapid innovation in ways to use ARPANET in computerized military command-and-control systems. DCA managers were also more concerned than ARPA had been about the security implications of unauthorized use. It was also "more serious than ARPA had been about preventing use of the network for 'frivolous' activities" (Abbate 1999, p. 136). These concerns were heightened when low-cost home computer systems began to appear in the late 1970s. The DCA instituted a new system of log-ins and passwords to ensure that only authorized terminal IMPs (TIPs) would have access to the network. Thus, while the transfer of operational responsibility helped to speed diffusion within the military, it initially slowed diffusion of civil applications.

A Common Language. In the late 1970s the DCA was confronted with a major decision that had a profound impact on the commercial development of the Internet. This decision involved the future of the Automatic Digital Network (AUTODIN)—the message switching network that the DCA had built for military use beginning in the early 1960s. An updated version, procured from Western Union, was scheduled to go into operation in 1979. The DCA had initially planned to dismantle ARPANET once AUTODIN II became operational. After further consideration the DCA decided that there was an important role in the DOD for a research-oriented network. DCA then decided to continue the research portion of ARPANET and set up a gateway to connect it to AUTODIN II. This meant that DCA had to confront the question of what to do about the TCP/IP, the new Internet protocol that Cerf and Kahn had developed for ARPANET. After considerable deliberation the ARPANET TCP/IP protocols were adopted in 1982 as the common language for the new Defense Data Network. After the Internet protocols were successfully tested in ARPANET, they became mandatory on the DOD networks.

After converting ARPANET to TCP/IP, the DCA and ARPA took several steps that set the stage for the development of a large-scale civilian Internet. The first step was to segregate the ARPANET military and academic functions. A decision was made in 1982, primarily on security considerations, to split ARPANET into a defense research network, still to be called ARPANET, and an operational military network, MILNET, which would be equipped with encryption devices and other security measures to protect the military functions. The second step involved support for the commercialization of the Internet technology. The DCA established a $20 million fund to subsidize the installation by computer manufacturers of TCP/IP on their machines: "All the major computer manufacturers took advantage of the opportunity and by 1990 TCP/IP was available for virtually every computer in the American market" (Abbate 1999, p. 43).

NSFNET. It was not until 1990, however, that ARPA was able to end its operational responsibility for ARPANET. In 1984 the National Science Foundation (NSF) established an Office of Advanced Scientific Computing. The mandate of the office was to organize a geographically dispersed set of new university-based supercomputing centers. An NSFNET was established as a "backbone" to link what eventually became sixteen centers. It was conceived as an internetwork rather than a single network. The "backbone" would be linked to the regional and local networks that NSF had earlier helped develop. An agreement was worked out with ARPA that, while the new NSFNET backbone was being built, NSFNET would use ARPANET as its backbone.

Abbate writes, "The NSF-ARPA interconnection opened up the Internet to nearly all universities in the United States" (1999, p. 193). It also created an opportunity for ARPA to end its operational responsibility for ARPANET. As early as 1987 the managers of ARPA's network program had decided that ARPANET had become obsolete and would have to be retired. The development of NSFNET opened up another option. Rather than dismantle ARPANET, they decided to "connect the ARPANET sites to the NSF's regional networks, and have the NSFNET take over as the backbone of the Internet." "When the NSFNET backbone was ready," Abbate explains, "it would be a simple matter to transfer the entire Internet community from the ARPANET to the NSFNET." Therefore, at the time "the ARPANET was decommissioned in February 1990, it marked the end of two decades of operation of the Internet by the military," and the "transition occurred so smoothly that few users were aware that the transition had taken place" (1999, pp. 194–195).[11]

Privatization. The first step in the privatization of the Internet began in 1983 when MILNET was split off from ARPANET.[12] As late as 1990, however, ARPANET was still operated by a government agency. Its use was restricted to nonprofit and educational institutions. The continuing ideologically motivated debates over federal subsidies to high-technology research and development convinced the NSF managers that the only politically feasible way to accommodate commercial users on the Internet would be to remove it entirely from government operation.

11. The growth of networks and internets grew dramatically between the middle and late 1980s. In 1986 Quarterman and Haskins published a 40-page article describing the population of specific networks and conferencing systems, and the interconnections between them. Four years later the same project required over 400 pages (Quarterman 1990, pp. 213–635).

12. For a more complete discussion of the issues of governance and regulation encountered during the process of privatization of the Internet, see Mowery and Simcoe (2003).

The process of privatization was largely completed by the mid-1990s. Two developments were necessary, however, to complete a global "network of networks"—the World Wide Web. One was the development of commercial Internet service providers (ISPs). ISPs are the primary servers and operators of the Internet infrastructure. In the United States and most other developed countries, ISPs are now private firms, but in many developing countries the ISP is a state or private telecommunications monopoly. The second was the development of an agency to administer the Internet name and address space. This issue was resolved in 1998 with the establishment of an international Corporation for Assigned Names and Numbers (ICANN). The primary functions of ICANN are to assign the unique IP numbers used to address a particular computer network and to oversee the operation of the Domain Name System (DNS) that maps these numbers onto domain names (Mowery and Simcoe 2003).

During the 1990s and into the first decade of the twenty-first century, Internet use in the United States and abroad experienced surprisingly rapid growth (Abbate 1999, pp. 195–200; Kenney 2003).[13] Mowery and Simcoe (2002) credit the deregulation of the U.S. telecommunications industry and the large size of the domestic market as important factors in more rapid adoption and diffusion of the Internet in the United States than in Western Europe and Japan. Greenstein (2000) has argued that Internet service providers needed to make only incremental changes in the technology and operating practices that were already pervasive in academic settings to be able to offer services to a wider variety of locations and users.

Perspective

From the late 1950s until the early 1990s, the DOD played a critical role in the invention and development of the Internet. Since the early 1960s ARPA has been the lead research agency in the DOD. During its first decade ARPA support enabled the development of the packet switching technology and the organization and demonstration of ARPANET to a largely skeptical computer science and communications research community. During its second decade ARPA created a new generation of technologies for packet radio, packet satellite, and internetworking. ARPANET went through a number of transformations, including a switch by the entire network community to TCP/IP, the splitting off of a MILNET from AR-

13. For more detailed discussion of the privatization process and the public and private institutional innovations that resulted in the development of the World Wide Web, see Abbate (1999, pp. 195–220) and Kogut (2003).

PANET, and the integration of ARPANET into the Internet. The next decade witnessed the absorption of ARPANET into the NSFNET and the first steps toward privatization of the Internet. Between the late 1950s and the early 1990s, DOD and NSF support for the development of what became the Internet amounted to close to $1 billion dollars (Langlois and Mowery 1996).

Abbate insists that, in the development of ARPANET, military concerns and values—survivability, flexibility, and high performance—dominated commercial goals, such as low cost, simplicity, or consumer appeal, from the time of Paul Baran's initial conceptual designs in the early 1960s to the transfer of ARPANET to civilian control in 1983 (Abbate 1999, pp. 5, 143–145). From the early vision of distributed computing by Licklider, to the collaborative approach to system design of Khan and Cerf, there was, however, a creative tension between the military imperatives and civilian applications, tension that was facilitated by the loosely structured administrative culture of ARPA during much of its history.

Since it was transferred to civilian control, users have largely lost sight of the contribution of military procurement to the development of the Internet. From the perspective of the commercial or individual user it is surely correct to assert that a critical date "by which to mark the explosion of the Internet onto the business and cultural scene is 1994, the year an easy-to-use Internet browser with secured transactions called Netscape was launched" (Kogut 2003, p. 2).[14]

Most of the scholars on whom I have drawn in this chapter have been reluctant to engage in counterfactual speculation about the economic impact of the development and adoption of the Internet. Norberg and O'Neill do enumerate the technical advances produced by ARPA and the research community that it supported: time sharing, wide-area networking, connections across networks, interactive graphics, distributed computing, natural language communications, and others (1996, p. 284). But they avoid a direct answer to the question, would it have happened anyway? They do insist, however, that no one could have expected the computer industry to finance and support innovation in computing on the scale that ARPA was able to do.

Robert Litan and Alice Rivlin have been less cautious. They suggest that the impacts of the Internet fall under four general interrelated categories: (1) reduction of the costs of the transactions involved in the production and distribution of goods

14. Netscape was the product of a research team led by Marc Andreessen that had been involved in the development of NSFNET at the National Center for Supercomputer Applications (NCSA) at the University of Illinois. The first version, named Mosaic, was released into the public domain in November 1993. It was the first system to include color images that could be used, like text words, as links. In 1994 Andreessen and his team left NCSA to develop a commercial version of Mosaic: Netscape.

and services, (2) increasing the efficiency of management in areas such as supply chain management, (3) making market relationships more transparent and competitive, and (4) increasing consumer choice, convenience, and satisfaction. Drawing on a series of commissioned industry studies, Litan and Rivlin estimate that the impact of the Internet on the annual rate of productivity growth could add something in the neighborhood of 0.25 to 0.5 percent to a baseline annual rate of growth in productivity in the U.S. economy during the early years of the twenty-first century (2001, pp. 19–22).

My own response to this question, although not inconsistent with that of Litan and Rivlin, is that, in the absence of military support for the development of computers in the 1940s and 1950s, and of microprocessors and integrated circuits in the 1960s and 1970s, realization of the Licklider vision would have been substantially delayed. Without the impetus from military procurement, I would not have been able to transmit this chapter to reviewers, or the book of which it is a part to the publisher, over the Internet in 2004. And it is doubtful that I would have been able to do so for at least another decade.

ACKNOWLEDGMENTS

I am indebted to Arthur Norberg and Jeffrey Yost for comments on an earlier version of this chapter.

REFERENCES

Abbate, J. 1999. *Inventing the Internet.* Cambridge, MA: MIT Press.

Baran, P. 1964. "On Distributed Communications Networks." *IEEE Transactions on Communications* (July): 1–9.

Cerf, V. G., and R. E. Kahn. 1974. "A Protocol for Packet Network Interconnection." *IEEE Transactions on Communications* COM-22 (May): 637–648.

Greenstein, S. M. 2000. "Computers and the INTERNET." *NBER Reporter* (Fall): 15–17.

Hafner, K., and M. Lyon. 1996. *Where Wizards Stay Up Late: The Origins of the Internet.* New York: Simon & Schuster.

Hughes, T. P. 1998. *Rescuing Prometheus.* New York: Random House.

Kenny, M. 2003. "The Growth and Development of the Internet in the United States." In *The Global Internet Economy,* ed. B. Kogut. Cambridge, MA: MIT Press, 69–108.

Kogut, B. 2003. *The Global Internet Economy.* Cambridge, MA: MIT Press.

Langlois, R., and D. C. Mowery. 1996. "The Federal Government Role in the Development of the U.S. Software Industry." In *The International Computer Software Industry: A Comparative Study of Industry Evolution and Structure,* ed. D. C. Mowery. New York: Oxford University Press, pp. 53–85.

Licklider, J.C.R. 1960. "Man Computer Symbiosis." *IRE Transactions on Human Factors in Electronics* 1: 4–11.

Litan, R. E., and A. M. Rivlin, eds. 2001. *The Economic Payoff from the Internet Revolution.* Washington, DC: Brookings Institution Press.

Mowery, D. C., and T. Simcoe. 2002. "Is the Internet a U.S. Invention?—An Economic and Technological History of Computer Networking." *Research Policy* 31: 1369–1287.

———. 2003. "Public and Private Participation in the Development and Governance of the Internet." Berkeley: University of California Haas School of Business. Mimeograph.

National Research Council. 1999. *Funding a Revolution: Government Support for Computing Research.* Washington, DC: National Academy Press.

Norberg, A. L., and J. E. O'Neill. 1992. *A History of the Information Techniques Office of the Defense Advanced Research Projects Agency.* Minneapolis: Charles Babbage Institute, University of Minnesota (October). Mimeograph.

———. 1996. *Transforming Computer Technology: Information Processing for the Pentagon, 1962–1968.* Baltimore: Johns Hopkins University Press.

Quarterman, J. S. 1990. *The Matrix: Computer Networks and Conferencing Systems Worldwide.* Bedford, MA: Digital Press.

Quarterman, J. S., and J. C. Haskins. 1986. "Notable Computer Networks." *Communications of the ACM* 29: 932–971.

Roland, R. 2002. *Strategic Computing: DARPA and the Quest for Machine Intelligence, 1983–1995.* Cambridge, MA: MIT Press.

Ruttan, V. W. 2003. *Social Science Knowledge and Economic Development: An Institutional Design Perspective.* Ann Arbor: University of Michigan Press.

7

The Space Industries

The launching of Sputnik, the first earth-orbiting satellite, by the Soviet Union (Union of Soviet Socialist Republics) on October 4, 1957, was interpreted by the press and regarded by the public as evidence of a much higher level of scientific and technical capacity in the Soviet Union than had been previously assumed. It challenged the perception that the United States was the unsurpassed leader in military, economic, and technological power. It has also been viewed as a defining event that led to dramatic scientific and technical advances in weather, communication, and earth-observing systems.

The launching of Sputnik II with a 4-meter-high cone equipped with a temperature-controlled cabin containing a dog on November 4, and the even larger and heavily instrumented Sputnik III on May 15, 1958, reinforced the perception of U.S. scientific and technical weakness and led to vigorous questioning of both civilian and military leadership by members of Congress.

My primary purpose in this chapter is to trace the development of weather, communication, and earth-observing satellites to their World War II and cold war origins. A secondary theme is the effect of defense-related security concerns and premature privatization policies on the achievement of commercial viability of the earth-observing satellite system.

Missiles and Satellites

The U.S. lag in the "space race" was less substantial than it appeared in 1957. Indeed, President Dwight D. Eisenhower and his immediate military and science advisors appeared not greatly alarmed by the apparent Soviet lead in launching an earth-orbiting satellite. The United States had been flying spy planes (the U-2) over the Soviet Union for more than a year and had initiated a program to develop sophisticated satellite observation capacity. Eisenhower saw the launching of the Soviet Sputnik as a useful precedent for an international "Open Skies" or "freedom of space" policy (Burrows 1986; Dickson 2001, pp. 2, 101).

The German Rocket Team

The U.S. capacity in missile and satellite science and technology in the early post–World War II period was, however, almost entirely based on the acquisition of the scientific and technical resources of the German rocket team, led by Werner von Braun, that developed and deployed the V-2 rocket against Great Britain during the war.[1] As the German defense was disintegrating, von Braun and his staff took steps to surrender to the U.S. Army. In the race between the United States and the Soviet Union to acquire German missile capacity, the U.S. Army was able to acquire most of the important German technical personnel and documents and much of the hardware, including almost all of the remaining V-2 rockets. The Soviet Union acquired most of the V-2 production team and some V-2 components (Logsdon 1970, p. 49; Dickson 2001, pp. 56–58).

In June 1945 the State and the War departments arranged for von Braun and more than one hundred of his scientific and technical personnel to be transferred to the United States to be "utilized in appropriate military purposes of the Army and Navy"—or, as anticipated by von Braun, to be "squeezed like lemons" and then sent back to Germany (Dickson 2001, p. 59). After a brief stay for debriefing at Wright Field near Dayton, Ohio, the team was transferred to an Army facility at Fort Bliss, near El Paso, in Texas. At Fort Bliss and the nearby White Sands facility, the team experimented with military rocketry, with an emphasis on altitude and payload. The status of the German rocket team remained uncertain, and somewhat controversial, until formal immigration processes were arranged (from Mexico) in 1949.

In late 1949 the Army moved the rocket team to the Redstone Arsenal in Huntsville, Alabama. At Redstone the pressures of the cold war dictated a shift from a

1. The problem that had to be solved to maintain a satellite in orbit was to balance the velocity of the satellite with the pull of the earth's gravity. The satellite must have enough velocity to offset the earth's gravitational pull, yet not enough to break away from the earth's gravity. Sir Isaac Newton had suggested that a satellite could be launched by a sufficiently powerful cannon located at the top of a high mountain. In 1898 the Russian mathematician Konstantin Eduardovich Tsiolkovsky demonstrated mathematically the feasibility of Newton's proposal. In the United States the engineer Robert Goddard demonstrated in a series of experiments conducted between 1914 and 1916 that a rocket could operate in space, and wrote a seminal article in 1919 on the problem of reaching high altitudes. His work was more influential in Germany and the Soviet Union than in the United States (Dickson 2001, pp. 28–48).

space vehicle orientation to an emphasis on the development of intercontinental ballistic missiles (ICBMs) capable of carrying a thermonuclear warhead. At Huntsville the von Braun team developed the Redstone, a super V-2 type rocket, and a variety of medium- and long-range ICBMs (Dickson 2001, pp. 62–64, 74–75).

During the late 1940s and early 1950s, there was substantial competition among the military services, and between proponents of military responsibility and proponents of civilian responsibility for missile development and space exploration (Day 1998, pp. 119–142). The proliferation of missile programs led to a series of inquiries and organizational initiatives that culminated in 1958 in the establishment of the Advanced Research Projects Agency (ARPA) in the office of the secretary of defense, in the creation of the position of director of defense research and engineering in the U.S. Department of Defense (DOD), and in the creation of a new independent agency, the National Aeronautics and Space Administration (NASA). NASA was charged with assuring leadership in space science and technology, including manned space flight; the launching of remote sensing, communication, and weather satellites; and aeronautical research and development (see chapter 3; York and Greb 1977; Bromberg 1999, pp. 16–44).

Project Vanguard

In April 1955 President Eisenhower approved plans to launch a satellite as part of an American contribution to the scientific activities associated with U.S. participation in the International Geophysical Year (IGY).[2] Shortly afterward he signed a National Security Council Directive that prohibited the U.S. IGY program from employing launch vehicles intended for military purposes" (Dickson 2001, p. 350). Eisenhower's objective was to emphasize the peaceful objectives of America's space program while avoiding the diversion of resources from the ballistic missile program (Logsdon 1970, p. 13).

2. The International Geophysical Year (IGY) was conceived by geophysicists associated with the International Council of Scientific Unions to demonstrate the possibility of peaceful cooperation among the world's scientific societies. Initial attention focused on the possibility of obtaining simultaneous measurements and observations of the earth and the upper atmosphere. Scientists from the United States, the Soviet Union, and other countries would cooperate to conduct atmospheric studies. The period July 1, 1957, to December 31, 1958, was selected partly because it would be the twenty-fifth anniversary of the second International Polar Year and partly because astronomers were anticipating an unusual level of solar activity during that period. As planning for the IGY progressed, both the United States and the Soviet Union announced intentions to launch satellites capable of conducting scientific measurements in the upper atmosphere (Dickson 2001, pp. 76–87). For a highly personal account of the events and personalities associated with the IGY, see Gavaghan (1999, pp. 1–36). I also draw on Gavaghan in the sections of this chapter dealing with weather satellites and space communication.

In the United States the IGY satellite program, termed *Project Vanguard,* was assigned to the Naval Research Laboratory, which had no involvement in the missile program. The objectives of the Vanguard program were "to develop and to procure a satellite launching vehicle; to place at least one satellite in orbit around the earth during IGY; to accomplish one scientific experiment; and to track the flight of the satellite" (Dickson 2001, p. 87).

Under pressure from the White House, a decision was made to commit the new and untested Vanguard rocket, Test Vehicle 3 (TV-3), to putting a satellite in orbit at Cape Canaveral on December 4, 1957. High winds and a frozen valve delayed the launch for two days. "Finally, writes Paul Dickson, "at precisely 11: 44:55 A.M. on Friday, December 6, 1957, with the whole world watching, the slender vehicle rose a few feet off the launch platform, shuddered slightly, buckled under its own weight, burst into flames, and collapsed. . . . Its tiny 3.2 pound payload, thrown free of the fire, rolled into the scrub brush and started beeping" (2001, pp. 156–157).

After a series of failures, the Vanguard II satellite was successfully launched on February 17, 1959. After two more Vanguard failures, the Vanguard III satellite, the last in the IGY series, was successfully launched on September 18, 1959. The Vanguard rocket was designed as a slim, efficient vehicle for scientific exploration, seventy-five feet in height. Vanguard "pioneered the use of solid-state devices, printed circuits, and the general principle of miniaturization. . . . It was the first to use solar cells to create power. . . . It gave proof that the Earth is slightly pear shaped" (Dickson 2001, pp. 179–184). Vanguard is expected to orbit the Earth for at least 1,000 more years.

Jupiter and Explorer

In February 1956 the Army Ballistic Missile Agency (ABMA) was established, under the command of Major General John B. Medaris, at the Redstone Arsenal near Huntsville, Alabama. It inherited von Braun's rocket team and the Redstone missile program that was already under way at Huntsville. In April von Braun suggested that the new Jupiter C missile, adapted from the Redstone, could be used to put a small satellite in orbit by January 1957, several months before the Vanguard was scheduled to be launched and, it was hoped, before the Soviet Union could put a satellite in orbit (Dickson 2001, p. 87).

The DOD, in conformity with the Eisenhower policy of separation of military and scientific missile developments, instructed Huntsville that "it was not to initiate any plans or preparations for using any part of the Redstone or the new Jupiter programs as the basis for an orbital launch vehicle" (Dickson 2001, p. 87). Medaris and von Braun made several insistent appeals, including a request that the Jupiter

C be designated as a backup to Vanguard. But even this request was rejected. Rejection did not, however, prevent ABMA from continuing to "bootleg" Jupiter tests over ranges and at altitudes that exceeded the limits mandated by the DOD.[3] In August 1957 Medaris placed two Jupiter C rockets in storage in case an authorization to launch a satellite was given (Dickson 2001, p. 92).

On January 31, 1958, following the aborted launching of the Vanguard TV-3, a backup Jupiter 3, renamed *Juno,* launched the first successful U.S. satellite into orbit. The satellite, renamed *Explorer I,* was an eighty-inch-long, 30.66-pound cylinder that contained 18.35 pounds of instruments. On March 5, a second Explorer was launched but never made it into orbit because of a fourth-stage rocket malfunction. On March 23 Explorer III, with an 18.83-pound payload, was successfully launched. The Explorer satellites were equipped with the same experimental package as Vanguard—a micrometeorite detector and a cosmic ray experiment—designed by James Van Allen of the University of Iowa. The detection by Explorer I and Explorer III of powerful radiation belts surrounding the earth, now referred to as the *Van Allen belts,* was the first major scientific discovery of the U.S. space program and the most important finding to emerge from the IGY (Dickson 2001, p. 182).

CORONA

At the time of the Sputnik "crisis" the U.S. Central Intelligence Agency (CIA), the Air Force, and several defense contractors (Lockheed, General Electric, and Fairchild), with the approval of President Eisenhower, were already working on a surveillance satellite project that was so secret that the existence of the program was known to only a few people (Greer 1995; Wheelon 1995; Day, Logsdon, and Latell 1998; Cloud and Clarke 1999a; Dickson 2001; Taubman 2003, pp. 193–355) For several months after its initiation, "by order of CIA Chief Allen Dulles all details were to be passed along verbally and there were to be no documents or written records" (Dickson 2001, p. 187).[4]

3. Medaris and von Braun complied with the DOD directive but evaded its intent (Dickson 2001, p. 89). On August 8, 1957, a Jupiter C booster, launched at Cape Canaveral, attained an altitude of 600 miles and a range of over 1,300 miles. Recovery of the nose cone marked the first recovery of an object from outer space and demonstrated that the nose cone could survive the intense heat of reentry (Dickson 2001, pp. 91–92).

4. A proposal for a system of surveillance, communication, and weather satellites was first made in 1946 in a classified report by RAND. The report envisioned a 1951 launch date (Dickson 2001, pp. 69–70). The CORONA project was the outgrowth of an Air Force spy satellite program: WS-117L. The program had developed so many security leaks that the Air Force pub-

The United States had initiated aerial reconnaissance over the Soviet Union in the summer of 1956, using the U-2 spy plane. It was anticipated that the Soviet Union would soon have radar sophisticated enough to track the flights. However, Soviet radar tracked the very first U-2 flight over Soviet territory. The Soviets registered a formal protest within days of the incident. The U-2 continued to range over much of the world for four years, but restricted its flights over the Soviet Union.[5] By the time the U-2 piloted by Francis Gary Powers was shot down near Sverdlovsk on May 1, 1960, the CORONA satellite was approaching operational readiness (Greer 1995, p. 3).

The CORONA program required the solution of a number of exceedingly difficult technical problems. Its first dozen launches were all failures. Of the first 30 missions, only 12 were productive (Greer 1995, p. 1). Camera design went through a series of revisions. Retrieving the film capsules involved the return of exposed film by parachute and midair recovery by aircraft. On August 18, 1960, the first fully successful CORONA satellite, Discoverer XIV, was launched: "This one satellite mission yielded photo coverage of a greater area than the total produced by all the U-2 missions over the Soviet Union (Greer 1995, p. 24).

The technical achievements of the CORONA project were summarized in 1973 by Kenneth Greer: "Its progress was marked by a number of notable firsts: the first to recover objects from orbit, the first to deliver intelligence information from a satellite, the first to produce stereoscopic satellite photography, the first to employ multiple reentry vehicles" (1995, p. 37). The advances in its earth-observing capacity were particularly noteworthy:

> From a single panoramic camera system having a design goal of 20 to 25 feet ground resolution and an orbital life of one day, to a twin camera panoramic system producing stereo-photography at the same ground resolution; then to a dual recovery system with an improvement in ground resolution to approximately 7 to 10 feet, and doubling the film payload; and finally to the J-3 system with a constant rotor camera, selectable exposure and filter controls,

---

licly canceled it in February 1958 (Dickson 2001, p. 187). It was covertly reactivated in 1958 (Greer 1995). President Eisenhower, disturbed by the lack of progress and frequent security leaks from the Air Force satellite program, directed that the CORONA project be developed and managed by the CIA with support from the Air Force and the new ARPA. From its inception in 1958 until it was phased out in 1972, the CORONA program was a continuing source of tension between the CIA and the Air Force. For a detailed history, see Taubman (2003, pp. 193–355).

5. The U-2 flights did contribute valuable information. In June 1957 a U-2 pilot flying over Kazakhstan detected the Soviet Union's most important missile test site. Photo interpretation enabled analysts to determine the size and power of Soviet rockets from burn marks and the configurations of the pads for exhaust gases (Dickson 2001, p. 100).

planned orbital life of 18 to 20 days, and yielding nadir resolution of 5–7 feet. (Greer 1995, p. 37)

The basic outline of the Eisenhower space policy as it evolved by the mid-1950s was characterized as "separation of civilian and military space projects, and a low estimate of the political significance of an (early) satellite launch" (Logsdon 1970, p. 13). In retrospect this policy reflected Eisenhower's concern that nothing interfere with the objective of negotiating an international "Open Skies" accord. This same perspective led him to insist that the civilian space effort, including planning for a lunar landing, be assigned to NASA when it was established in 1958. In 1960 the von Braun team at Huntsville was formally transferred to NASA, thus ending Army plans for the development of manned space flight (Logsdon 1970, pp. 29–62).[6]

By the early 1960s the potential strategic and economic contributions of the several space programs were beginning to become apparent. The program of ABMA, motivated by the energetic bureaucratic entrepreneurship of von Braun, had set in motion the development of the technology that led to the NASA manned space flight program. Project Vanguard, initiated under the auspices of the IGY program, laid the groundwork for NASA initiatives in space communications technology and space science. The several Air Force surveillance projects advanced the earth-observing technology that became the foundation for much of modern geographic information systems (GIS).[7]

Weather Satellites

The first of the NASA applications programs to reach the operational stage was the Meteorological Satellite Program (Metsat).[8] The first weather satellite program,

6. I have discussed the establishment and organization of NASA in chapter 3. For a more detailed account, see Bromberg (1999, pp. 15–44).

7. I do not attempt in this chapter to discuss the more recent developments in military space programs or the technology of the manned or unmanned space flight programs other than to refer to the launch facilities and delivery vehicles developed and employed in the space programs. See Johnson (2002b) for the U.S. ICBM program, and Johnson (2002a, pp. 115–153) for project Apollo, the U.S. manned space flight program. For a very useful account of the development of U.S. and Soviet military satellites, see Stares (1987, pp. 8–44). In 1961 James E. Webb was appointed administrator of NASA. Under his leadership NASA was transformed from a small politically vulnerable agency to one well on its way to a manned moon landing when he retired in 1968. Although Webb was committed to advancing a broad program that included space science, communications, and earth-observing satellites, there was never any question that he regarded manned space flight as the centerpiece of the NASA program (Lambright 1995). For a more skeptical perspective, see McDougall (1985, pp. 361–388).

8. This section is adapted from Mack (1990, pp. 19–23). For an excellent history of the tech-

Project TIROS (Television and Infra-Red Observation Satellite), was initiated as part of an ABMA reconnaissance satellite program. The project was transferred to NASA in 1959. The U.S. Weather Bureau, located in the Department of Commerce, had little influence, or even contact, with the program to develop a weather satellite until after the launch of the first experimental TIROS I satellite in 1960.

The Weather Bureau became involved when NASA asked it to evaluate the utility of the data being returned by the TIROS I satellite. Bureau meteorologists were enthusiastic about the value of the data. Almost immediately, the Weather Bureau began to produce and distribute cloud cover maps. By 1962 the Weather Bureau was providing daily transmission of cloud cover maps to meteorologists.

As NASA began to plan for a more sophisticated Nimbus series of proto-operational satellites, a series of interagency disputes emerged with the two primary users of the weather data: the U.S. Weather Bureau and the U.S. Air Force. The Weather Bureau found the data provided by TIROS, operated by the DOD and NASA, satisfactory for its purpose.[9] The coarse resolution imagery provided by TIROS was not a problem for the Weather Bureau. It was concerned that the Nimbus data would be considerably more expensive and that there might be a gap in data availability before Nimbus became operational. The U.S. Air Force, still resentful that the TIROS project had been transferred to NASA, was also concerned about the possibility of a gap between the TIROS and Nimbus projects, which would leave it without a storm warning system. It was also concerned with the strategic implication of releasing finer-resolution imagery. The interagency conflict was resolved by an agreement with the Weather Bureau that NASA would continue to maintain TIROS as an operational system while going ahead with Nimbus as a NASA research and development project (Lambright 1994).[10] The failure of NASA to give early consideration to potential user needs, in addition to its own research and development agenda, emerged as a more serious obstacle to the transition from experimental to operational programs in its communication and earth-observing satellite programs.

nical development of the series of weather satellites, see Williamson (1995). For a more detailed account of events, politics, and personalities involved, see Lambright (1994) and Gavaghan (1998, pp. 129–168).

9. Since the late 1970s the NASA weather satellites have been used to study vegetation and land cover; to monitor crop failures, range conditions, deforestation, and desertification; and to develop famine warning systems (Morain 1998, p. 31).

10. The series of TIROS satellites became operational in 1966. They were renamed the NOAA Polar Orbiting Environmental Satellites in 1970.

Space Communications

An important unanticipated consequence of the space race, initiated by the successful launching of Sputnik by the Soviet Union, was the rapid development of communications satellites by the United States (Gavaghan 1998). A proposal for a high-altitude, geostationary, earth-orbiting satellite for telecommunications was first suggested in 1945 by Arthur C. Clarke, a member of the British Interplanetary Society and then serving as a Royal Air Force radar officer (Clarke 1945). In 1952 John R. Pierce of Bell Laboratories, writing without knowledge of Clarke's article, suggested several promising satellite configurations including both low-altitude and high-altitude orbits (Durant 1984). But it was President John F. Kennedy's sense of urgency about demonstrating American superiority in space that gave impetus to the technical and institutional innovations that resulted in a commercially viable space communications industry.

In 1961 President Kennedy issued a "Policy Statement on Communications Satellites" (Kennedy 1961). The statement recognized the potential economic value of satellites in providing communications services and recommended (1) government policy for the conduct and coordination of research and development, (2) implementation by the public sector, and (3) an international effort in which all nations would be invited to participate. Kennedy's statement was followed by passage of the Communications Satellite Act of 1962, by formation of the Communications Satellite Corporation (COMSAT) in 1963, and by organization of an International Telecommunications Satellite Consortium (INTELSAT) in 1964.

Experimentation

During the late 1950s and early 1960s, a number of active communications satellite tests were conducted. The first satellite, labeled *SCORE*, was built by the DOD and launched into orbit on an Atlas rocket in 1958. It was designed to receive a message from earth, store it on tape, and transmit the message to ground: "SCORE transmitted what became known as the first 'voice from space' when it was used to transmit a Christmas message to the world from President Eisenhower. The battery power for its transmitter died on New Year's Eve" (Edelson 1984, p. 45). In October 1960 the DOD launched Courier, a much more sophisticated communication satellite. It was solar-cell powered and had four receivers and transmitters. Courier functioned perfectly for eighteen days and then failed because of a command-system fault. SCORE and Courier demonstrated "that delicate and complex electronic equipment could be made to survive the trauma of rocket launch and function in orbit" (Edelson 1984, p. 45).

In the early 1960s the American Telephone and Telegraph (AT&T) Bell Laboratories used its own corporate resources to develop Telstar, a more advanced communication satellite. It was launched by NASA on a Thor-Delta rocket into an elliptical orbit on July 10, 1962. The Telstar surface was covered by solar cells and was circled by an antenna belt. On the day Telstar was launched, it relayed live television transmitted from the United States to England and France. Telstar II was launched into a higher altitude in May 1963. It demonstrated "that a medium-altitude commercial satellite system was feasible and, to many, that such a system was preferable" (Edelson 1984, p. 46).

At the time President Kennedy made his 1961 space policy statement committing the United States to the development of a space communications system, the technology needed to achieve the goals he set out still had to be demonstrated. No test of voice or video transmission from a satellite had yet been conducted. No data on reliable operation of electronic deices or rotating mechanisms in space had yet been produced. There was considerable doubt as to the acceptability of the delay in voice transmission from high-altitude geosynchronous satellites for commercial service (Treubel and Steinmueller 1982; Edelson 1984, p. 41).

Echo. Beginning in the late 1950s, NASA initiated a program designed to resolve these issues.[11] It first initiated a program to develop and launch a series of passive satellites.[12] The first ECHO satellite, launched from Cape Canaveral into orbit in August 1960, was a metal and plastic balloon with a diameter of ninety-eight feet: "It provided the first live, two way voice communication via satellite. . . . Within a week of its launching the first transoceanic satellite message, transmitted from Bell Laboratories in New Jersey, was received in Paris by an earth station of the Centre Nationale d'Etudes de Telecommunications" (Edelson 1984, p. 44). A somewhat larger ECHO II was launched by NASA in January 1964. There was no expectation that the ECHO satellites would ever develop into operational systems. Their purpose was to test signal propagation and to develop transmission techniques.

Relay. Even before Telstar was launched for AT&T, NASA had contracted with Radio Corporation of America (RCA) to develop a second medium-altitude communications satellite. The satellite, RELAY I, was launched in December 1962; RELAY II, in January 1964. The RELAY satellites had the capacity to transmit

11. For an authoritative description of the construction, technology, and capacity of the series of experimental satellites launched between the late 1950s and the mid-1960s, see Jaffe (1966). For a more colorful history, see Tedeschi (1989).

12. Work by NASA on active and geosynchronous design was delayed until the early 1960s, pending resolution of territorial disputes with the DOD (Treubel and Steinmueller 1982).

one live television program or twelve simultaneous two-way telephone conversations. Both satellites transmitted live television around the world for several years. The importance of the RELAY program to the development of communications satellites was that (1) a successful satellite design was implemented, thus paving the way for incremental improvement and commercial adaptation, and (2) ground station problems of satellite tracking and data acquisition were solved (Treubel and Steinmueller 1982, p. 273). The one critical limitation of the medium-orbit Telstar and RELAY satellites was the necessity of tracking the moving satellites.

Syncom. At the time Telstar and RELAY were being launched, NASA had awarded a contract to Hughes Aircraft for the construction of a geostationary synchronous satellite. In 1958–1959 two Hughes engineers, Harold Rosen and David Williams, had developed the concept of a spin-stabilized spacecraft to be placed in a geostationary orbit. The proposal had been initially rejected by the Army and regarded skeptically by AT&T. It received a more positive response from NASA (Bromberg 1999, pp. 45–56).

The SYNCOM I satellite was launched into an elliptical orbit with use of a Thor-Delta rocket in February 1963 but failed to survive the apogee motor firing.[13] The success of SYNCOM II, launched in July 1963, and of SYNCOM III, launched in August 1964, demonstrated two substantial technical and economic advantages of the geostationary orbit for almost all satellite communications applications. One was that global coverage could be achieved by only three such satellites. In contrast many dozens of the lower-orbit Telstar or RELAY satellites would have been required to provide even, continuous transatlantic coverage. A second was that geostationary satellites allow earth station antennas to be fixed, thus eliminating the need for expensive tracking equipment and movable antennas (Lovell 1973, pp. 40–50; Edelson 1984, p. 18; Cohen and Noll 1991, p. 150).

Commercialization

As launching of the several communications satellites proceeded, the need for an operational entity to implement the system became increasingly apparent to the communications industry, the Congress, and the administration. Senator Robert Kerr (D-OK) proposed the creation of a private communications satellite system.

13. The apogee is "the point in an elliptical satellite orbit which is farthest from the surface of the Earth. Geosynchronous satellites, which maintain circular orbits around the Earth are first launched into highly elliptical orbits with apogees of 22,237 miles. When the communications satellite reaches the appropriate apogee, a rocket motor is fired to place the satellite into its permanent circular orbit of 22,237 miles" (Pelton 1991, p. 206).

The Kennedy administration's bill proposed setting up a corporation in which ownership would be divided between the private commercial carriers and the public, and would be governed by a board consisting of six representatives each of the common carriers and the public, and three appointed by the president. A third proposal was introduced by Senator Estes Kefauver (D-TN), which called for an entirely government-owned satellite system. After a lengthy and often acrimonious debate, the Communications Satellite Act, similar in most respects to the administration proposal, was enacted on August 31, 1963 (Pelton 1974; Colino 1984).

While the creation of COMSAT was being negotiated, the Kennedy administration was also moving to establish a global communications satellite system. European governments organized a European Conference of Post and Telecommunication Administrations (CEPT). A series of meetings between the U.S. State Department and COMSAT officials and representatives of CEPT in 1963 and 1964 resulted in the formation of INTELSAT in 1964. It was also agreed that COMSAT would serve as the manager of INTELSAT. Ownership would be based on the signatories' contributions to the capital costs for the design and establishment of facilities. Initially the United States would receive 61 percent; Western Europe, 30.5 percent; and Canada, Japan, and Australia, 8.5 percent. The initial percentages would be modified as other countries joined.[14] An INTELSAT Interim Communication Satellite Committee (ICSC) was established as a liaison body with COMSAT to monitor the design, development, construction, maintenance, and operation of the systems satellites.

The early NASA satellite series—ECHO, RELAY, and SYNCOM—were designed to demonstrate and advance the technology of using satellites for routing relay and data transmission, and for intercontinental communications (Smith 1976, p. 233). SYNCOM III served as a prototype for the first generation of commercial satellites that began operation under the auspices of INTELSAT. At the time ICSC was formed, a contract with Hughes for the first INTELSAT satellite had already been placed by COMSAT.

INTELSAT I, popularly known as *Early Bird*, was launched in April 1965. It had a capacity of 480 telephone channels. At that time the largest transatlantic cable operated by AT&T had a capacity of 256 channels (Pelton 1974, p. 65). INTELSAT II, also built by Hughes Aircraft, was launched in October 1966 to provide satellite coverage of the Pacific basin. Global coverage was achieved in July 1969 when INTELSAT III, built by Thompson, Ramo, Woodridge (TRW), was

14. For an exceedingly detailed technical and institutional history of the development of the INTELSAT system, see Science Policy Research Division, Congressional Research Service (1983, pp. 11–218). See also Pelton (1974); Smith (1976, pp. 121–155); and Colino (1984).

placed in orbit over the Pacific Ocean. Technical improvements involving greater power and energy transmission efficiency "made it possible for the INTELSAT III system of satellites operating over the three ocean regions to provide 1200 voice circuits, four television channels, or a combination of telephone and television."[15] In its second week of operation, it "transmitted live man's first lunar landing to virtually every corner of the globe, setting a new record global television audience of 500 million people" (Colino 1984, p. 79).

In the mid-1960s congressional budget concerns led NASA to begin to reevaluate its satellite commitments. In 1965 NASA informed the Congress that it had decided to discontinue launching communications satellites and proposed an Advanced Technology Satellite, later termed *Applications Technology Satellite* (ATS), program. The shift was also consistent with NASA's announced policy of focusing on technology that would not otherwise be developed by the private sector. Under the ATS program five experimental satellites of widely varied design, built to NASA specifications, were constructed by Hughes Aircraft. These were launched between late 1966 and August 1969. Only two of these satellites, ATS-1 and ATS-5, functioned completely as planned. Nevertheless, the program has been evaluated as highly successful in developing some of the critical technologies employed in the first four INTELSAT satellites.

Planning for two additional ATS satellites, the ATS-F and ATS-G, began in early 1966. Contracts were awarded, after considerable controversy, first to General Electric and then, after formal protest, to Fairchild-Hiller in September 1970. Before the ATS-F was launched in 1974, scheduling delays and cost escalation led to the canceling of the contract for the ATS-8. Cohen and Noll insist, however, that the ATS-F program "was not a failure" (1991, p. 163).[16] It led to a number of technical advances in areas such as antenna technology, frequency levels, signal

15. The design and construction of INTELSAT III involved important procedural and contractual innovations by INTELSAT. It was the first satellite series contract award that went through a full international request for proposal process. The contract was awarded to TRW. For a detailed technical description of each series of satellites through INTELSAT, see Podraczky and Pelton (1984). COMSAT Laboratories, established in 1967 on the model of Bell Telephone Laboratories, played an exceedingly important role in monitoring contract awards and development work by contractors (Tedeschi 1989).

16. Cohen and Noll, who are generally critical of federal funding of technology development programs, note, "While the ATS program was generally regarded as an enormous success, it was canceled when NASA was preparing to pursue the next logical step. This contrasts with other programs, in which efforts appear to have continued far past the point of diminishing marginal returns" (1991, p. 167). The most obvious answer is that when President Nixon sought to cut all space programs, NASA placed lower priority on further development of communication satellites at the expense of its manned space programs.

transmission power, and ground station design. The ATS-6 satellite was an important breakthrough in achieving high orbital broadcast power and dramatically reducing the size and cost of earth-based receivers. It also indicated the possibility of using satellites for new types of satellite services, such as health and education in remote locations (Treubel and Steinmueller 1982, p. 276).[17]

In 1972 the Federal Communications Commission (FCC) authorized qualified private firms to launch and operate domestic satellite systems. The first privately operated U.S. domestic satellite, Westar I, was launched at Cape Kennedy by NASA for Western Union on April 13, 1974 (Smith 1976, pp. 176–185). In 1973 NASA, still under budgetary pressure, announced that the private sector had reached a level of scientific and technical maturity that the agency could completely phase out its research and development (R&D) on communications satellites. Strong support for continuation of the ATS by the leadership of the Senate Aeronautical and Space Science Committee and by private-sector user organizations led NASA to concede that, if the private sector failed to respond adequately to maintain U.S. leadership in communications satellite technology, it might be forced to reevaluate it phase-out decision (Smith 1976, pp. 230–245).

During the 1970s several developed-country members became increasingly uncomfortable with the monopoly arrangement for management of INTELSAT by the U.S. COMSAT.[18] After several abortive efforts a European Space Agency (ESA) was formed in 1972 to coordinate all European space R&D, with the objective of developing an independent launch and satellite capacity. In 1979 France used its Ariane rocket to put a communication satellite into orbit from the ESA launch site in French Guiana. France then formed a "private" company, Ariane-

17. Treubel and Steinmueller, using exceedingly conservative assumptions, including that the entire benefit of NASA investment in satellite communication technology from the initiation of the ECHO program in the late 1950s through the SYNCOM program accrued to savings in the U.S.-Europe submarket between 1965–1970, estimated direct resource savings of at least $49.5 million. They comment that they doubted they could find many cases in which the first direct impact of the baseline technological level of a major innovation covered one third of the costs within six years. They note that a more complete accounting would include (1) the share of resource savings resulting from base-period knowledge and technology development that occurred after 1970, and (2) the share of the value to society derived from the satisfaction of previously unfulfilled communication needs. They go on to argue that the NASA space communication experience demonstrated that public support for technology development may be justified even beyond the stage of early communication (Treubel and Steinmueller 1982, pp. 281–284).

18. Before the early 1980s a non-Communist country interested in launching a communication satellite had only one choice: NASA. To uphold its commitment to INTELSAT, the United States insisted that any other satellites it launched could be used only for experimental purposes. For the evolution of U.S policy with respect to international competition in space launch services, see Smith (1992).

space, which successfully marketed launch services for communication satellites by INTELSAT, several Latin American and Arab countries, and a few private U.S. firms (McDougall 1985, pp. 424–429).[19]

By the late 1980s access to international satellite services for voice communications, data communications, and video services was broadly available. In addition to INTELSAT, there were regional providers such as Eutelsat, PanAmSat, ARABSAT, and AsiaSat, as well numerous domestic systems that provided some regional coverage, such as AUSSAT (Australia), Palapa (Indonesia), INSAT (India), and Brasilsat.

When the Reagan administration took office in 1981, it attempted to move rapidly toward privatization of launch services. Commercialization of satellite communication services had been successful. Why not commercialization of launch activities? The only aerospace firm to seriously respond to a 1982 NASA request for proposals was General Dynamics. Its bid presumed that it would be able to rent government launch activities at a price that reflected incremental cost, that it would be able to purchase spare parts at a favorable price, and that NASA would continue to conduct research and development work on launch vehicles and provide launch facilities. Privatization of commercial satellite launch services was actively pursued by the Reagan and Bush administrations (Bromberg, 1999, pp. 128–131).[20]

As early as the mid-1980s, questions were being raised about the future technical, economic, and political viability of the COMSAT-INTELSAT system (Demac 1986). In the early 1970s RAND had estimated that a low-orbit COMSAT system could add capacity at a cost of about $8,500 per channel per year—about one third of the cost of transoceanic cables (Smith 1976, pp. 60–76). A decade later a RAND study indicated that the introduction of fiber-optical cables had sharply reduced the cost of terrestrial communications systems and was leading to

19. During the 1970s, NASA spending fell to 36 percent of its Apollo peak in constant dollars. "Policy straitjackets and funding cutbacks stymied the United States in exploiting its own technology" (McDougall 1985, p. 429). It was also stymied by the reluctance of the recently privatized communication satellite industry to undertake the investments in technology development and testing that were anticipated by the proponents of privatization (Cohen and Noll 1991, pp. 161–177).

20. As of the early 2000s, NASA was responsible for conducting launches of the space shuttle, and the Air Force was responsible for launches of military and intelligence satellites. All other U.S. launches are conducted by private-sector companies. Facilities for launching private satellites are provided by NASA (Smith 2003). Virtually every rocket used to launch commercial payloads, whether U.S. or foreign, was until at least the late 1980s still based on ICBM technology (U.S. Congress, Office of Technology Assessment 1990).

excess telecommunication (satellite plus terrestrial) capacity (Johnson 1987; Johnson 1988).[21]

But by the late 1990s telecommunications analysts were hailing a satellite renaissance. Wireless services and satellites were experiencing record growth. A global telecommunications system was emerging that would include high-altitude, geosynchronous, earth-orbiting satellites (GEO); low earth-orbiting satellites (LEO); and a "rich but confused digital mixture of fiber, coaxial cable, terrestrial wireless and satellite services carrying everything from voice to broadband multimedia and video services" (Pelton 1998, p. 82). Most of these systems were expected to be privately operated and directly competitive with the COMSAT-INTELSAT system. By the early 2000s it was apparent that the anticipated boom in demand for commercial satellite services had failed to materialize (Pasztor 2003; Sequeo and Pasztor 2003; Smith 2003).

Earth-Observing Systems

In this section I discuss the role of military procurement in the emergence of earth-observing systems. It will be useful, however, to first review the role of military procurement in the post–World War II development of the field of GIS.

Geographic Information Systems

The lineage of GIS traces back to the origins of cartography and to the development of map overlay systems in the nineteenth century.[22] Modern GIS owes its devel-

21. Edwin Layton writes, "In the mid-1960s scientists at ITT's Standard Telecommunications Laboratory in England suggested that light waves could be guided by glass to where they were needed. By 1970 scientists at Corning Glass had made the idea work. Hair-thin pieces of silica glass could bend easily to serve as "waveguides" for light waves. In an optical fiber, light is funneled in one end, is repeatedly reflected at a low, critical angle off the walls of the fiber, and emerges at the same angle at the other end—as if it had been placed in a pipeline. This property of optical fiber holds true no matter how many turns and twists the fiber makes along its length. Fiber optic cable provides transmission opportunities far beyond and far cheaper than any conventional medium" (1986, p. 22).

22. For a history of GIS, see Tomlinson, Wilkins, and Marble (1976); Coppock and Rhind (1991); and Foresman (1998). H. H. Tomlinson, who developed the Canadian Geographic Information System (CGIS) in the mid-1960s with support from the Canadian Department of Agriculture, is regarded as the "father of GIS." Other early contributors include Howard Fisher of the Harvard Laboratory of Computer Graphics, and David Bickmorat of the Experimental Cartography Unit in the United Kingdom (Coppock and Rhind 1991). For a perspective on the technical and intellectual issues that confront the development of geographical information science, a perspective that still retains its currency, see Goodchild (1992).

opment and spectacular growth in the last quarter of the twentieth century to two technical developments: the computer and the earth-orbiting satellite. The computer brought about a transition from making maps by hand to the use of digital technology to produce three-dimensional maps (Clarke and Cloud 2000).[23] The CORONA program, discussed earlier in this chapter, played an important role in both developments.

John Cloud and Keith Clarke have insisted that the impact of the CORONA program was so pervasive that it has been difficult to identify "any significant Geographic Information Sciences technologies, applications, or data sets which do not have a primary or secondary origin in collaboration with the secret assets of the military and intelligence institutions" (Cloud and Clarke 1999a).[24] They also insist that the significance of the technology developed in connection with the CORONA program was not fully understood, even by many of the scientists involved in the development of GIS, until the declassification of the program in late 1995 (Cloud and Clarke 1999a).

Cloud and Clarke also suggest that the development of GIS represents a paradigm for military and civilian collaboration in "dual use" institutions, technologies and applications that they term the *shuttered box*.[25] The relationship between the World Geodetic System (WGS) (classified) and the U.S. Geological Survey (USGS; unclassified) is an example. "Civilian researchers typically got unrestricted access

23. The social and political implications of these technical changes have been controversial in the field of geography. Some students of GIS have been critical of what they consider the excessive emphasis by Cloud and Clarke on the impact of the CORONA program on the development of GIS. GIS has been criticized as a "new imperialist geography." See, for example, Pickles (1995) and Sheppard (1995).

24. For greater detail, see Cloud (2000, pp. 112–217). Cloud reviews the impact of military procurement, particularly the CORONA program, on the development of the geodetic sciences. During the 1950s and 1960s, the Air Force supported the establishment and operation of the Institute of Geodesy, Photogrammetry and Cartography (IGPC), at Ohio State University. The IGPC was the first advanced degree–granting institution in the geodetic sciences in the United States. It also operated a program of nondegree advanced training and workshops. Participants in both the degree and nondegree programs were drawn largely from the U.S. and foreign military services. The overarching research objective was the development of an integrated world geodetic data system and an earth model. The earth model enabled ICBMs to be accurately targeted, and ICBM launch sites to be detected by space satellites. The success of the research program led to its demise. In 1973 the geodetic sciences at Ohio State were merged into the Department of Civil Engineering.

25. Cloud and Clarke write, "On one side is the classified world populated by those with clearance. On the other side is the open world of civilian science. The Shuttered Box works in this manner: by coordinating the opening and closing of shutters on all sides, the view through the box is precluded at all times—there is an absolute separation between entities on either side of the box—by opening and closing shutters in tandem, materials and people can pass securely *in either direction* back and forth through the box. That which can and has passed through the box includes: (1) funding; (2) people and their experience; (3) tools and techniques; (4) findings and related data; and (5) knowledge and science" (1999b, p. 45).

only to the degraded version of WGS released publicly by the USGS. Even when civilian agencies were given access to undergraded versions of CORONA photography for use in cartographic applications, the origins of the imagery and the data source were completely disguised" (Cloud and Clarke 1999b, p. 49). The shuttered box metaphor was appropriate in this case, because the civilian world already possessed resources relevant to the success of CORONA, cartographic institutions and geodetic theory, for example, while CORONA had access to technology that was beyond the capacity of any civilian organization and most government agencies.

Landsat

During the early 1960s NASA scientists and engineers were initiating studies and experiments to develop technology to monitor earth resources from space. When NASA first initiated the Earth Resources Technology program in 1964, the program was viewed more as an exercise that would enable NASA to put more people into space than as a program with specific resource-monitoring or management objectives. Relatively little effort was made to assess the needs of agencies such as the U.S. Weather Bureau, the USGS, and the U.S. Department of Agriculture (Mack 1990, pp. 45–55).[26]

Among the several agencies, the USGS of the Department of the Interior was the most aggressive in attempting to take advantage of satellite technology. In September 1966 Secretary Stewart Udall announced that the Department of the Interior was planning its own program of Earth Observing Satellites (EROS) with a possible first launch in 1969. In retrospect it appears that the Interior initiative was taken in an effort to accelerate the NASA program and to generate an incentive to give greater attention to users' needs, as opposed to its own internal priorities.[27]

The NASA response to external interest in the development of applications was slowed by greater interest on the part of NASA program managers in the development of more advanced instrumentation, such as sensors, than in application.

26. The studies of the Landsat satellite system by the Science Policy Research division of the Congressional Research Service at the Library of Congress (1983, pp. 219–293) and by Pamela E. Mack (1990) remain the most comprehensive studies of the program. For a very useful technical history, see Irons (2000). For reviews of U.S. remote-sensing policies, see O'Connell and Hilgenberg (2001, pp. 139–163), and Rowberg (2002). Although the name of the program to develop an Earth Resources Technology Satellite was not changed to Landsat until 1972, the year Landsat I was launched, I follow Mack in referring to the program as Landsat from its start.

27. Documents that have been declassified since the study by Mack (1990) suggest that the national security agencies thought that any public disclosure of information based on remote sensing from satellites would be harmful to national security. For a more detailed account of the tension and the cooperation between the defense and civilian agencies over Landsat applications for civilian purposes, see Cloud (2000, pp. 217–277).

Even more important was the opposition to the development of earth-observing technology for civilian application by the security agencies and the Bureau of the Budget. Technology that had been developed in the CORONA program was unavailable for use in Landsat. The issue was so sensitive that "the Bureau of the Budget refused to allow NASA even to list mapping as an objective of Landsat" (Mack 1990, p. 75). It was not until May 1969 that NASA finally obtained enough funding to put out requests for proposals for design of the earth resource satellites that would meet the requirements specified by its own studies and those of potential users, particularly EROS. General Electric was chosen as prime contractor in mid-1970 (Mack 1990, pp. 94–106).

The first Landsat satellite was successfully launched on July 23, 1972. It was immediately apparent that "the political future of earth resource satellites depended on the development of an enthusiastic group of users who would act as a constituency for the program" (Mack 1990, p. 122). Even before the first Landsat satellite was launched, operational uses were being planned by the USGS. One of the most important was the identification of those geological formations where petroleum and mineral deposits might be located. Other unanticipated applications included the identification of the location and quality of water resources. The success of the Interior Department in developing applications led to selection of EROS as the distribution center for Landsat data. In other federal agencies, such as the Department of Agriculture, and in state and local agencies, operational use occurred more slowly because the methods for interpretation of Landsat data to meet agency needs had to be developed, or the establishment of agency capacity to use the data required specialized training (Mack 1990, pp. 196–211).

By the time Landsat 4 was launched in 1982, it had become the "workhorse" for environmental research dealing with earth surface monitoring and analysis. It was apparent that government support for Landsat could no longer be rationalized on the basis that it was still an experimental program. Yet the Office of Management and Budget (OMB) refused to allow NASA to employ a public-goods rationale as part of its justification for increased funding.[28] Questions of ownership and management of the system had to be addressed. In the case of communications satellites, the initiative had been taken by the telecommunications industry and

28. The Office of Management and Budget (OMB) "required a whole series of cost-benefit studies, in which NASA had to justify Landsat not on the grounds of new (economic) benefits that would result but on the grounds of how much money the satellite project would save the government by replacing old ways of doing things" (Mack 1998, p. 229). This failure to consider the public-goods dimension of government research and development was consistent with my own experience when I was told by a young OMB examiner, "It is not the business of the government to make a profit!"

COMSAT. After substantial debate President Jimmy Carter transferred Landsat to the National Oceanic and Atmospheric Administration (NOAA), which continued to administer the successful weather satellite program. Carter also instructed NOAA to develop a plan for the privatization of Landsat. The Land Remote-Sensing Commercialization Act of 1984 authorized commercial operation of the Landsat satellites under contract with, and subsidized by, NOAA.

In 1985 Landsat was acquired by the Earth Observing Satellite Corporation (EOSAT), a joint venture of Hughes Aircraft and RCA Astro-Electronics. EOSAT was charged with operating Landsat 4 and 5 under contract with NOAA, completing the privatization of Landsat services, and launching several more advanced satellites (Landsat 6 and 7). EOSAT immediately raised the price of Landsat images tenfold—from $400 to $4,000 per image (David 2004, p. 12). The effect was to preclude purchase of Landsat images by academic and independent users and to limit the market primarily to government and commercial users. Even after privatization, an ideologically burdened debate about government support for the Landsat system continued into the early 1990s. It was increasingly recognized that full commercial viability would not be achieved until well into the next century. The problem of meeting the competing demands of the several research communities, national security agencies, and civil and commercial users was an obstacle to achieving scale economies. Remote-sensing data, much of it superior to that being produced by the older Landsat satellites, became available from the French SPOT system. The European Space Agency, Canada, Japan, and India were planning earth-observing satellite launches. Congress delayed funding of Landsat 7, which it had earlier authorized (see table 7.1 Radzanowski 1991).

By the early 1990s it was clear that the corporate owner of Landsat was unable, or unwilling, to devote to the program the resources that would be required to achieve technical or economic viability. The Land Remote-Sensing Policy Act of 1992 repealed the commercialization act of 1984. After a series of rather convoluted interagency negotiations, NASA was assigned responsibility for building and launching Landsat 7, and the USGS was assigned responsibility for postlaunch satellite and ground systems operations. Landsat 7 was built for NASA by Lockheed Martin and successfully launched on April 15, 1999. As of the mid-1990s there remained considerable skepticism that "the United States would maintain leadership in either technology for civilian earth observation or a robust earth observing operational program" (Mack 1998, p. 235).[29]

29. James Irons provides a more positive perspective while recognizing the problem that had been created by premature commercialization efforts: "The Landsat Program claims the longest record of global observations from space. The record is, however, terribly fragmented as the data

Table 7.1. Chronology of Landsat and Landsat-Like Launches, 1972–2007

Year	Platform (Nationality)	Year	Platform (Nationality)
1972	Landsat-1 (United Sates)	1996	ADEOS (Japan)
1975	Landsat-2 (United Sates)	1996	PRIRODA (Germany/Russia)
1978	Landsat-3 (United States)	1997	IRS-1D (India)
1982	Landsat-4 (United States)	1998	CBERS (China/Brazil)
1984	Landsat-5 (United States)	1998	SPOT-4 (France)
1986	SPOT-1 (France)	1998	Landsat-7 (United States)
1988	RESURS-01 (Russia)	1998	EOS AM-1 (United States/Japan)
1988	IRS-1A (India)	1998	IRS-P5 (India)
1990	SPOT-2 (France)	1999	Resource 21 (United States)
1991	IRS-1B (Japan)	2000	IRS-2A (India)
1992	JERS-1 (Japan)	2002	ALOS (Japan)
1993	Landsat-6 (United States)	2002	SPOT-5A (France)
1993	SPOT-3 (France)	2004	IRS-2B (India)
1993	IRS-P1 (India)	2004	SPOT-5B (France)
1994	IRS-P2 (India)	2004	ALOS-A1 (Japan)
1994	RESURS-02 (Russia)	2007	ALOS-A2 (Japan)
1995	IRS-1C (India)		

Source: Stanley A. Morain, 1998, "A Brief History of Remote Sensing Applications with Emphasis on Landsat," in *People and Pixels: Linking Remote Sensing and Social Science*, ed. D. Liverman, E. F. Moran, R. Rindfuss, and P. C. Stern (Washington, DC: National Academy Press), 45.

Note: For the full names of platforms and sensors, see S. A. Morain and A. M. Budge, eds., "Earth Observing Platforms and Sensors," vol. 2, in *Manual of Remote Sensing*, ed.-in-chief P. Ryerson, 3rd ed., vol. 2 (Bethesda, MD: American Society for Photogrammetry and Remote Sensing), CD-ROM. Reprinted with permission of the American Society for Photogrammetry and Remote Sensing.

As of late 2003 Landsat 7 was approaching the end of its operationally useful life. Planning for Landsat 8 was delayed. Negotiations with a private contractor for the development of a satellite system for a Landsat Data Continuity Mission had been terminated. The scientific user community had expressed concern that privatization would again have the effect of pricing academic research out of the market. Even greater alarm was expressed about the potential loss of continuity in more than thirty years of Landsat earth resource observations (Gower et al. 2003).

are distributed amongst the ground station operators. . . . Landsat data for long-term or large-scale investigations becomes exceedingly difficult to access and exploit. . . . This situation creates a data processing overhead that is daunting and has hindered multiple scene analysis" (2000, pp. 5–6).

Global Positioning

The impact of advances in remote sensing on the development of civil technology and institutions was severely limited until well into the 1990s by the secrecy associated with earth-observing technology and information. The situation has changed rapidly, however. By the early 1990s the U.S. Air Force was supporting development of the initial Navstar global positioning constellation of twenty-four orbiting satellites, at least four of which were above the local horizon anywhere on earth for twenty-four hours a day. They emitted two sets of signals, which allowed users to calculate their precise location anywhere on earth. One was an encrypted code for use by the U.S. military and selected allies, and the second was an open signal for civilian use. During the 1990s the Global Positioning System (GPS) became increasingly viewed as a global public utility (box 7.1).

At the same time, substantial concerns were being raised about the future of GPS. The degrading of civilian signals, termed "selective availability," had become a constraint on commercial development (National Academy of Public Administration 1995). The termination of selective availability after an executive order by President Clinton in May 2000 resulted in substantial improvements in accuracy and stimulated a new burst in commercial demand. There remained considerable skepticism about the commitment of the DOD to making the necessary technical and financial investments necessary to enhance GPS capacity consistent with the growth in commercial demand.

The European Union, motivated by both strategic and commercial concerns, made an official decision in March 2003 to challenge the monopoly status of GPS by building the Galileo, an independent European satellite constellation. The Galileo system of thirty state-of-the-art satellites was projected to be operational by 2008. Arguments by the DOD that Galileo would imperil U.S. and North Atlantic Treaty Organisation (NATO) security interests had not, at least by mid-2003, succeeded in altering European Union policy (Braunschvig, Garwin, and Maxwell 2003).

At the time this chapter was written, it seemed clear that during the next several years additional countries and private firms would have high-resolution remote-sensing satellites in orbit. Many of these, particularly those operated by government agencies, will have dual use: military and commercial. But most of these will be selling an appreciable share of their imagery in the open market. And it is doubtful that any government, or even consortium of governments, will have the capacity to put meaningful constraints on access to the highest-quality images (Florini and Dehqanzada 2001).

Box 7.1. Origins of the Global Positioning System

The Global Positioning System (GPS), a satellite-based system enabling remarkably precise pinpointing of one's location on Earth, evolved from postwar work on atomic clocks to test aspects of general relativity theory. Their possible value for navigation was recognized by the military, which provided years of "patient federal capital" to mature the technology. The military's primary interest in what was to become GPS was to improve the delivery of tactical weapons and to reverse the proliferation of costly new navigation systems. Early in its development GPS was recognized as a potential dual-use technology.

In 1972 several military programs involved in what was to become GPS coalesced when the Air Force was given responsibility for developing a navigation system for all military services, as well as civilian users. Concurrently, technologies essential to GPS, including the CORONA satellites and microelectronics, also were being developed. Experimental GPS satellites were launched in 1978. The initial objectives were to improve navigation for military aircraft and ships, and to increase the accuracy of the weapons that they carried. Use of GPS for navigational purposes on civilian aircraft was approved in September 1983 after the Soviet downing of Korean Airlines Flight 007 over Soviet airspace (Braunschvig, Garwin, and Maxwell 2003).

By the mid-1990s the Navstar system consisted of a constellation of twenty-four earth-orbiting satellites, each carrying up to four atomic clocks that provided timing and ranging signals. A GPS receiver decoded the signals to determine and display its latitude, longitude, and altitude. Differential GPS is the most widely used method for augmenting basic GPS signals and now yields centimeter accuracies over distances of several kilometers. This translated into an incredible array of applications, such as demonstrating new systems for landing aircraft in bad weather, site-specific fertilizing and planting of fields, monitoring train and truck locations, tracking and cleaning up oil spills, and the siting and location of industrial and commercial facilities. In 1995 the global commercial GPS market was estimated at $2.3 billion, was already larger than the military market, and was projected to reach upward of $10 billion by 2000.

Source: Adapted from Committee on Criteria for Federal Support of Research and Development (1995, box II.2, p. 49). See also National Academy of Public Administration (1995); Rip and Hasik (2002); and Zhang, Wang, and Wang (2002).

Perspective

The initial decision by the U.S. government to put a satellite into orbit was based entirely on military and strategic considerations. Even before Sputnik was launched in 1957, the United States was attempting to confront a crisis in its capacity to monitor Soviet military capacity and deployment. The Eisenhower administration wanted to establish an international "freedom of space" regime that would legitimize its intention to launch a global system of reconnaissance. This interest coincided with that of the international geophysics community in designating 1957–1958 as an IGY. At the time the first satellites were launched, neither the United States nor the Soviet Union had given significant attention to the potential nonmilitary uses of applications satellites—for weather forecasting, communication, earth observation, or geographic positioning.

A question that has not been adequately addressed in the literature is, why did nonmilitary and intelligence applications of satellite communication technology achieve commercial viability so much more rapidly than satellite earth-observing and geographic positioning systems? One answer is that in the case of satellite communication technology the delivery systems for telephone and television transmission were already in place. Commercial markets developed rapidly. Rapid diffusion of satellite communications technology was a response to a demand or a need that could not be met efficiently, or in some cases at all, by existing land-based technology. In addition, latent commercial demand was reinforced by political motivation, in the wake of the Sputnik launch by the Soviet Union, to publicly demonstrate U.S. scientific and technical capacity for peaceful application of space technology.

In the case of earth-observing technology, there was only a limited preexisting commercial market for the services that the new technology could provide. The initial demand for the data that could be provided by GPS was primarily for military application. The initial nonmilitary sources of demand for the data that could be provided by Landsat were primarily public-sector resource-management and planning agencies. Furthermore, the development and diffusion of the most advanced earth-observing technology and data were constrained by national security considerations.

Development of civil applications was caught up in ideological debates that led to premature privatization, and in concerns about the security implications of civil release of earth-observing data. It is also apparent in retrospect that reluctance to consider the public-good aspects and insistence on premature commercialization have been a serious constraint on efforts to sustain the United States' initial preeminence in earth-observing systems.

In spite of these differences, there was one essential common element that made the several systems possible. That element was the development, first by the military services and later by NASA, of the ICBM launch capacity necessary to place and maintain in orbit the "voice-from-the-sky" and the "eye-in-the-sky" satellites. The vehicles that were capable of launching and placing the satellites into orbit became available only because of the enormous prior investment by the military services and NASA. These expenditures were induced and precipitated by the cold war tensions between the United States and the Soviet Union.

In the absence of the development of launch vehicle capacity by the military services and NASA, it is difficult to believe that even AT&T, arguably the world leader in commercial telecommunications and electronics in the mid-1950s, would have made the investments that led to the development of the space communication and earth-observing industries.

ACKNOWLEDGMENTS

I am indebted to Keith C. Clarke, Eileen Galloway, Michael F. Goodchild, Pamela Mack, Joseph Pelton, Erik Sheppard, Marcia Smith, and B. L. Turner for comments on an earlier version of this chapter.

REFERENCES

Braunschvig, D., R. L. Garwin, and J. C. Maxwell. 2003. "Space Diplomacy." *Foreign Affairs* 82 (July–August): 156–174.

Bromberg, J. L. 1999. *NASA and the Space Industry*. Baltimore: Johns Hopkins University Press.

Burrows, W. E. 1986. *Deep Black: Space Espionage and National Security*. New York: Random House.

Clarke, A. C. 1945. "Extraterrestrial Relays." *Wireless World* (October): 305–308.

Clarke, K. C., and J. G. Cloud. 2000. "On the Origins of Analytical Cartography." *Geography and Geographic Information Science* 27: 195–204.

Cloud, J. 2000. "Hidden in Plain Sight: The Clandestine Geography of the Cold War." University of California, Santa Barbara. Ph.D. dissertation.

Cloud, J., and K. C. Clarke. 1999a. "The Fubini Hypothesis: The Other History of Geographic Information Science." University of California, Santa Barbara, Department of Geography. Mimeograph.

———. 1999b. "Through a Shutter Darkly: The Tangled Relationships Between Civilian, Military and Intelligence Remote Sensing in the Early U.S. Space Program." In *Secrecy and Knowledge Production*, ed. J. Reppy. Ithaca, NY: Cornell University Peace Studies Program, Occasional Paper 23 (October), 35–55.

Cohen, L. R., and R. G. Noll. 1991. "The Applications Technology Satellite Program." In *The Technology Pork Barrel*, ed. L. R. Cohen and R. G. Noll. Washington, DC: Brookings Institution Press, 149–178.

Colino, R. R. 1984. "The INTELSAT System: An Overview." In *The Intelsat Global*

Satellite System, ed. J. Alper and J. N. Pelton. New York: American Institute of Aeronautics and Astronautics, 55–94.

Committee on Criteria for Federal Support of Research and Development. 1995. *Allocating Federal Funds for Science and Technology*. Washington, DC: National Academy Press (the Press Report).

Coppock, J. T., and D. H. Rhind. 1991. "The History of GIS." In *Geographical Information Systems: Principles and Applications*, ed. D. J. McGuire, M. Goodchild, and D. W. Rhind. New York: Wiley, 21–43.

David, P. A. 2004. "Can 'Open Science' Be Protected from the Evolving Regime of IPR Protections?" *Journal of Institutional and Theoretical Economics* 160: 9–34.

Day, D. A. 1998. "A Strategy for Reconnaissance: Dwight D. Eisenhower and Freedom of Space." In *Eye in the Sky: The Story of the CORONA Spy Satellites*, ed. D. A. Day, J. M. Logsdon, and B. Latell. Washington, DC: Smithsonian Institution Press, 119–142.

Day, D. A., J. M. Logsdon, and B. Latell, eds. 1998. *Eye in the Sky: The Story of the CORONA Spy Satellites*. Washington, DC: Smithsonian Institution Press.

Demac, D. A. 1986. *Tracing New Orbits: Cooperation and Competition in Global Satellite Development*. New York: Columbia University Press.

Dickson, P. 2001. *Sputnik: The Shock of the Century*. New York: Walker Publishing.

Durant, F. C., III. 1984. "Prologue and Pioneers." In *The Intelsat Global Satellite System*, ed. J. Alper and J. N. Pelton. New York: American Institute of Aeronautics and Astronautics, 21–37.

Edelson, B. I. 1984. "The Experimental Years." In *The Intelsat Global Satellite System*, ed. J. Alper and J. N. Pelton. New York: American Institute of Aeronautics and Astronautics, 39–53.

Florini, A. M., and Y. A. Dehqanzada. 2001. "The Global Politics of Commercial Observation Satellites." In *Commercial Observation Satellites: At the Leading Edge of Global Transparency*, ed. J. C. Baker, N. M. O'Connell, and R. A. Williamson. Santa Monica, CA: RAND, 433–448.

Foresman, T. W., ed. 1998. *The History of Geographic Information Systems: Perspectives from the Pioneers*. Upper Saddle River, NJ: Prentice Hall PTR.

Gavaghan, H. 1998. *Something New under the Sun: Satellites and the Beginning of the Space Age*. New York: Springer-Verlag.

Goodchild, M. F. 1992. "Geographical Information Science." *International Journal of Information Sciences* 6: 31–45.

Gower, S. N., A.F.H. Goetz, J. B. Schott, J. J. Thome, J.R.G. Townshend, S. L. Ustin, and C. E. Woodcock. 2003. "Landsat Continuity Mission and the Long-Term Landsat Mission: Open Letter to Science Community Landsat Date Users." College Park: University of Maryland, Department of Geography. Mimeograph.

Greer, K. E. 1995. "CORONA." *Studies in Intelligence*, Suppl. 17 (1973): 1–37. Repr. in *CORONA: America's First Satellite Program*, ed. K. C. Ruffner. Washington, DC: Central Intelligence Agency, Center for the Study of Intelligence, 3–39.

Irons, J. R. 2000. "The Present and Future of the Landsat Program." Greenbelt, MD: Biospheric Sciences Branch, NASA Goddard Space Flight Center. Mimeograph.

Jaffe, L. 1966. *Communications in Space*. New York: Holt, Rinehart and Winston.

Johnson, L. L. 1987. *Issues in International Telecommunications: Government Regulation of Comsat.* Santa Monica, CA: RAND.

———. 1988. *The Future of INTELSAT in a Competitive Environment.* Santa Monica, CA: RAND.

Johnson, S. B. 2002a. *The Secret of Apollo: Systems Management and American and European Space Programs.* Baltimore: Johns Hopkins University Press.

———. 2002b. *The United States Air Force and the Culture of Innovation, 1945–1965.* Washington, DC: Air Force History and Museum Program.

Kennedy, J. F. 1961. "Communications Satellite Policy." Washington, DC: Department of State Bulletin 45. Repr. as "Statement by the President on Communications Satellite Policy." In *Papers of President John F. Kennedy 1961.* Washington, DC: U.S. Government Printing Office, 529–531.

Lambright, W. H. 1994. "The Political Construction of Space Satellite Technology." *Science Technology and Human Values* 19: 47–69.

———. 1995. *Powering Apollo: James E. Webb of NASA.* Baltimore: Johns Hopkins University Press.

Logsdon, J. M. 1970. *The Decision to Go to the Moon: Project Apollo and the National Interest.* Cambridge, MA: MIT Press.

Lovell, B. 1973. *The Origins and International Economics of Space Exploration.* Edinburgh, U.K.: Edinburgh University Press.

Mack, P. E. 1990. *Viewing the Earth: The Social Construction of the Landsat Satellite System.* Cambridge, MA: MIT Press.

———. 1998. "Landsat and the Rise of Earth Resource Monitoring." In *From Engineering Science to Big Science: The NACA and NASA Collier Trophy Research Project Winners,* ed. P. E. Mack. Washington, DC: National Aeronautics and Space Administration, 235–250.

McDougall, W. A. 1985. *The Heavens and the Earth: A Political History of the Space Age.* New York: Basic Books.

Morain, S. A. 1998. "A Brief History of Remote Sensing Applications with Emphasis on Landsat." In *People and Pixels: Linking Remotes Sensing and Social Science,* ed. D. Liverman, E. Moran, R. Rindfuss, and P. C. Stern. Washington, DC: National Academy Press, 23–50.

Morain, S. A., and A.M. Budge, eds. "Earth Observing Platforms and Sensors." In *Manual of Remote Sensing,* ed.-in-chief P. Ryerson, 3rd ed., vol. 2. Bethesda, MD: American Society for Photogrammetry and Remote Sensing. CD-ROM.

National Academy of Public Administration. 1995. *The Global Positioning System: Charting the Future.* Washington, DC: National Academy of Public Administration.

O'Connell, K., and G. H. Hilgenberg. 2001. "U.S. Remote Sensing Programs and Policies." In *Commercial Observation Satellites: At the Leading Edge of Global Transparency,* ed. J. C. Baker, K. M. O'Donnell, and R. A. Williamson. Santa Monica, CA: RAND, 139–164.

Pasztor, A. 2003. "Satellite Bonanza Is a Fizzle." *Wall Street Journal,* 9 September, B4.

Pelton, J. N. 1974. *Global Communications Satellite Policy: INTELSAT, Politics and Functionalism.* Mt. Airy, MD: Lomond Books.

———. 1991. *The "How To" of Satellite Communications.* Sonoma, CA: Design.

———. 1998. "Telecommunications for the 21st Century." *Scientific American* (April): 80–85.

Pickles, J., ed. 1995. *Ground Truth: The Social Implications of Geographic Information Systems.* New York: Guilford Press.

Podraczky, E., and J. Pelton. 1984. "INTELSAT Satellites." In *The Intelsat Global Satellite System,* ed. J. Alper and J. N. Pelton. New York: American Institute of Aeronautics and Astronautics, 95–133.

Radzanowski, D. P. 1991. *The Future of Land Remote Sensing Satellite System (Landsat).* Washington, DC: Congressional Research Service, Report 91–685 SPR.

Rip, M. R., and J. M. Hasik. 2002. *The Precision Revolution: GPS and the Future of Aerial Warfare.* Annapolis, MD: Naval Institute Press.

Rowberg, R. E. 2002. *Commercial Remote Sensing by Satellite: Status and Issues.* Washington, DC: Library of Congress, Congressional Research Service, RL 31218.

Science Policy Research Division, Congressional Research Service. 1983. *United States Civilian Space Programs, Vol. 2: Applications Satellites.* Washington, DC: U.S. Government Printing Office.

Sequeo, A. M., and A. Pasztor. 2003. "U.S. to Pay $538 Million to Keep 2 Firms in Rocket-Launch Field." *Wall Street Journal,* 12 February, A6.

Sheppard, E. 1995. "GIS and Society: Toward a Research Agenda." *Cartography and Geographic Information Systems* 22: 5–16.

Smith, D. D. 1976. *Communications via Satellite: A Vision in Retrospect.* Boston: Sijthoff.

Smith, M. S. 1992. *Space Launch Services—The Competitive Playing Field: A Primer.* Washington, DC: Library of Congress, Congressional Research Service, 92-933-SPR.

———. 2003. *Space Launch Vehicles: Government Activities, Commercial Competition, and Satellite Export.* Washington, DC: Library of Congress, Congressional Research Service.

Stares, P. B. 1987. *Space and National Security.* Washington, DC: Brookings Institution.

Taubman, P. 2003. *Secret Empire: Eisenhower, the CIA, and the Hidden Story of America's Space Espionage.* New York: Simon & Schuster.

Tedeschi, A. M. 1989. *Live via Satellite: The Story of COMSAT and the Technology That Changed World Communication.* Washington, DC: Acropolis Books.

Tomlinson, R. F., H. W. Wilkins, and D. F. Marble. 1976. *Computer Handling of Geographical Data.* Paris: UNESCO.

Treubel, M., and E. Steinmueller. 1982. "Government Policy, Innovation and Economic Growth: Lessons from a Study of Satellite Communications." *Research Policy* 11: 271–287.

U.S. Congress, Office of Technology Assessment. 1900. *Access to Space: The Future of U.S. Space Transportation Systems.* Washington, DC: U.S. Government Printing Office.

Wheelon, A. D. 1995. "Lifting the Veil on CORONA." *Space Policy* 11: 249–260.

Williamson, M. 1995. "And Now the Weather: Early Development of the Meteorological Satellite." *Newcomen Society for the Study of Engineering and Technology Transactions* 66: 53–76.

York, H. F., and G. A. Greb. 1977. "Military Research and Development: A Postwar History." *Bulletin of the Atomic Scientists* (January): 13–27.

Zhang, N., M. Wang, and N. Wang. 2002. "Precision Agriculture—A Worldwide Overview." *Computers and Electronics in Agriculture* 36: 113–132.

8

Is War Necessary?

In an earlier book, *Technology, Growth, and Development* (2001), I concluded that government-sponsored research and technology development played an important role in the development of almost every general-purpose technology in which the United States was internationally competitive. In that book I discussed the role of public-sector research and technology development (R&D), and of other forms of public support, on the development of the agricultural, electric light and power, chemical, computer and semiconductor, and automobile industries.

I noted that between the early 1950s and the late 1970s defense and defense-related R&D accounted for about two thirds of public-sector support for R&D and that government support exceeded private support for R&D every year until the late 1970s (figure 8.1; Ruttan 2001, pp. 547–532). During this same period military spending has experienced several cycles but has not risen in real (inflation-corrected) terms since the mid-1960s. As a share of gross domestic product and as a percentage of federal government spending, it has declined steadily (figure 8.2). I did not, however, examine in depth the role of defense and defense-related R&D and procurement on technology development in *Technology, Growth, and Development.*

In this book I have reviewed the development of six general-purpose technologies in which defense or defense-related procurement has played an important role in technology development. In each of these cases, commercial development would have been substantially delayed without the stimulus from military and defense-related procurement. In one case, nuclear power, it is doubtful the technology would have been developed at all in the absence of military procurement.[1]

1. In appendix 1, I list the critical dates in the development of the several general-purpose technologies that I discuss in this book, from the date of initial conception or invention, through development for military and commercial applications. It is at these stages that military and defense-related R&D and procurement has had the greatest impact. In appendix 2, I construct a counter factual history of the interrelated computer-microprocessor-Internet technologies.

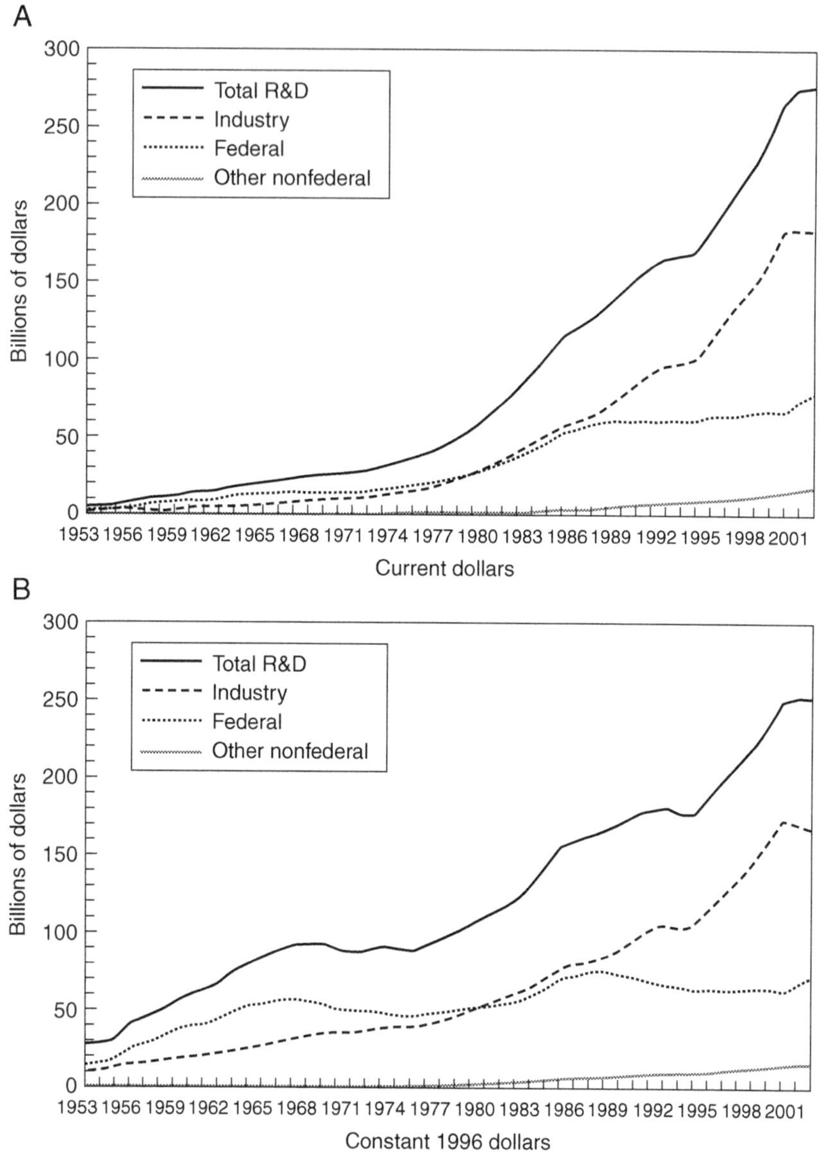

Figure 8.1. National research and development funding by source, 1953–2001: (A) in current dollars, (B) in constant 1996 dollars. *Source:* National Research Council (2005, p. 19).

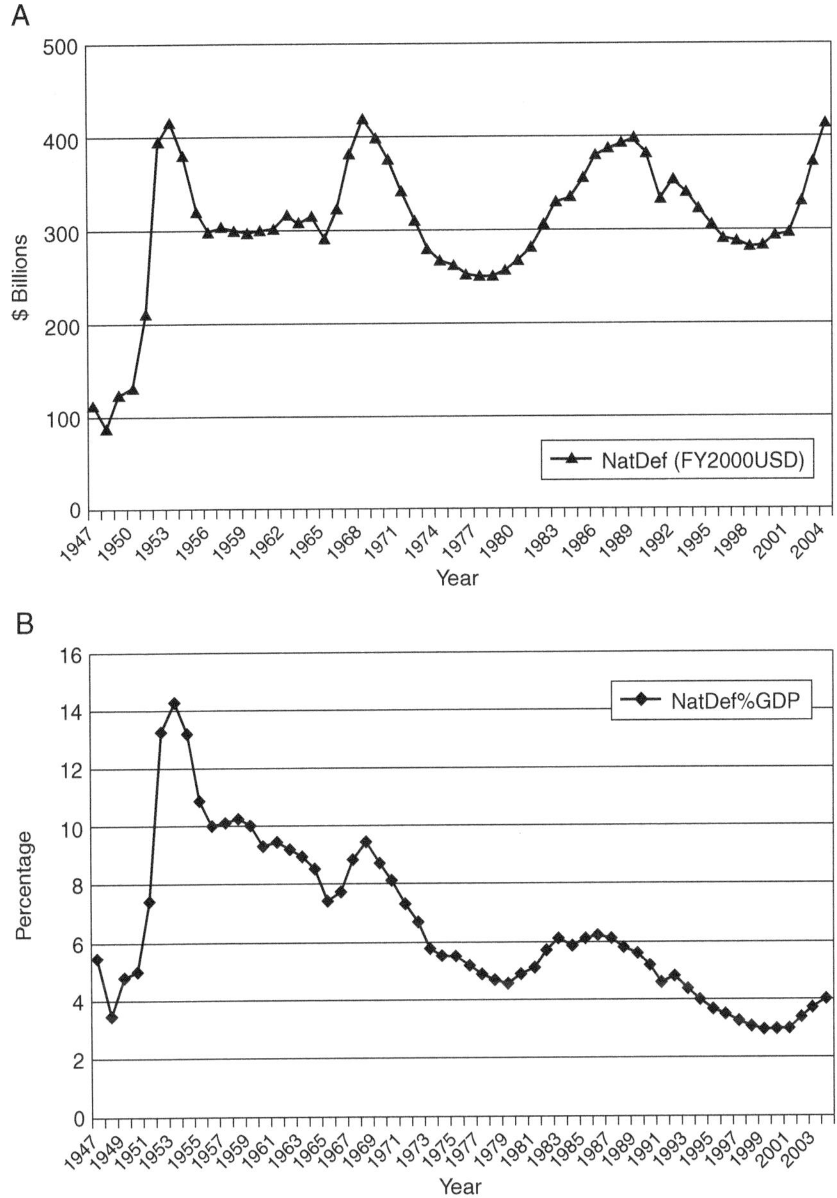

Figure 8.2. U.S. military spending, fiscal year 1947 through fiscal year 2004: (A) constant (2000) dollars; (B) as a percentage of gross domestic product in current dollars; (C) as a percentage of federal government outlays in current dollars. *Source:* Gold (2005). The original data source is Office of Management and Budget, Executive Office of the President of the United States (2004, table 6.1).

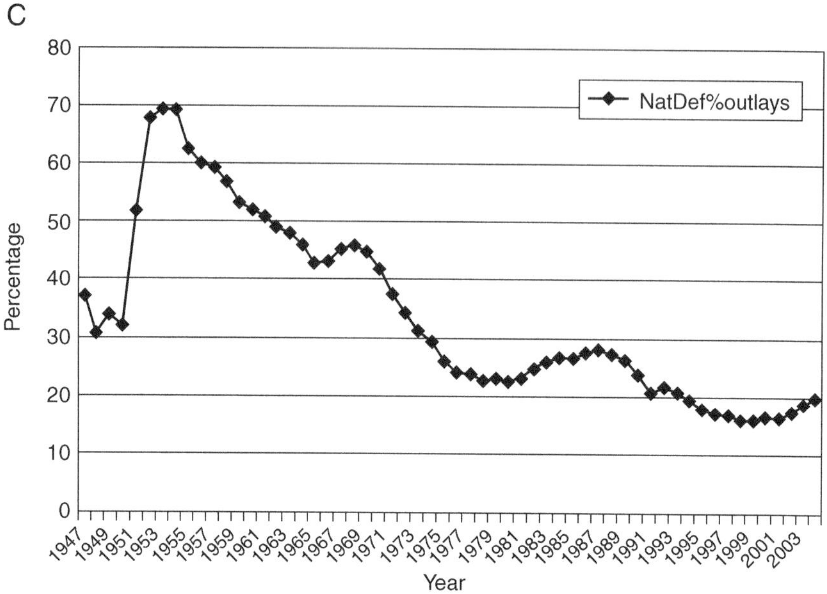

Figure 8.2. *(continued)*

I do not argue that the massive military and defense-related R&D and procurement programs that I have reviewed in earlier chapters can be adequately evaluated in terms of their impact on commercial technology development. They must be evaluated primarily in terms of their impact on national security.[2] Nor do I argue that the technology spin-off from expenditures on defense and defense-related research, technology development, and procurement is an efficient way to advance commercial technology development. Much of military and defense-related technology has been inordinately expensive. With rare exceptions benefit-cost calculations have not been carried out.

I do insist, however, that the American, and the global, technological landscape in which we live today would be vastly different in the absence of military and defense-related contributions to commercial technology development.

In this final chapter I return to several issues that have emerged in the process of writing the earlier chapters. The first is the issue of technological maturity. The second is whether changes in the structure of the U.S. economy and in the military industrial base preclude military and defense-related procurement from continuing

2. For a review of the literature on the impact of defense and defense-related spending on economic stability and growth, see Gold (1990). Gold concludes that "defense spending does not provide a consistent explanation of the aggregate performance of the U.S. economy" (p. 3).

to play an important role as a source of new general-purpose commercial technologies. The third is whether a major war or threat of major war is necessary to induce the United States to mobilize the R&D resources necessary to advance new general-purpose technologies.

Technological Maturity

A major deficiency in the induced, evolutionary, and path-dependency literature on technical innovation (see chapter 1) is inadequate attention to the problem of technological maturity. After experiencing rapid or even explosive development along an initial trajectory, the older general-purpose technologies, discussed in earlier chapters, have often experienced a period of technological maturity or stagnation. In some cases renewed development has occurred along a new technological trajectory. In other cases alternative paths of technology development have not yet become apparent. In a classic case that has received a great deal of attention in the economic history literature, the electric power industry, measurable impact of a revolutionary new technology on economic growth occurred only as the technology approached technological maturity (David 1990; chapter 4).

There is a modest literature on innovation and product cycles.[3] A formal model of the innovation cycle was first advanced by Robert Evenson and Yoav Kislev (1975). They traced the history of sugarcane technology development through three innovation cycles beginning in the middle of the nineteenth century. As each cycle reached maturity, the productivity of research effort directed to the development of new technical varieties declined. In the first two cases, however, advances in basic biological knowledge opened up new opportunities for a new round of technology development.

A similar model was employed by Cowan and Foray in an analysis of the complementarities between military and commercial technology. They argued that when R&D is performed primarily with military applications in mind, scientific and technical effort will be focused on identifying the part of the distribution that is most directly related to military application; when commercial applications are the primary focus, technical effort will focus on development of the most profitable varieties: "Military and civilian R&D will explore different (though perhaps over-

3. In the mid-1960s Raymond Vernon (1966) advanced a product cycle model to interpret the initial invention, modification, and adoption of consumer durables in the United States (or other high-income countries) and the later transfer of production to low-wage economies as the technology matured. See also Vernon (1979) and Ruttan (2001, pp. 163–166).

lapping) parts of the technological distribution" (Cowan and Foray 1995, p. 861). Failure to advance along either the military or commercial part of the trajectory can, in their model, dampen technical advances in the other. In contrast, advances in defense-related technology can induce technology development in the commercial sector, and feedback from advances in the commercial sector can induce technology development in the defense sector.

There is no implication that because an industry becomes mature and is no longer a dynamic source of growth in its country of origin it cannot continue to be a source of modest output and productivity gains as it makes the transition from a leading to a sustaining sector. Similarly, as a result of international technology transfer, it may become a dynamic source of growth in less technically advanced countries.[4]

Mass production in the automobile industry achieved technological maturity in the United States during the 1930s. It became a leading sector in several European countries and Japan during the 1950s. During the 1960s a series of technical and institutional innovations characterized as "just in time" and "total quality control" assembly were being implemented in Japan. By the 1970s they had begun to displace the traditional "Fordist" mass production system in the United States and other developed countries in automobile production (Ruttan 2001, pp. 445–452; chapter 2). Lower labor productivity in the production of wide-bodied jets at Boeing than at Airbus in the early 2000s was a consequence of failure by Boeing to move beyond traditional mass production technology and organization.

Aircraft propulsion is an example of an industry in which a mature technological trajectory was followed rapidly by transition to a new technological trajectory. Piston-propeller aircraft propulsion achieved technological maturity in the late 1930s. The scientific and technical foundations for a transition to a jet propulsion trajectory were well under way by the late 1940s. In the absence of military support for R&D during World War II and military procurement during the Korean War, the transition to jet commercial aircraft propulsion would have occurred much more slowly (chapter 3).

The technology of electric power generation from coal-fired power plants reached technological maturity between the late 1950s and the early 1960s with boiler-turbine units in the 1,000-megawatt range.[5] In the United States the tran-

4. The concept of "leading sector" played a central role in the growth-stage theories of economic development (Rostow 1956, 1960).

5. "The technical design frontier was limited by the ability of boilers to withstand high temperature and pressure. The frontier was pushed out by incremental advances, particularly in metallurgy, involving the development of high-temperature alloys. Most of the shift to higher temperatures and to reheat cycles was completed in the 1948–57 decade with little change thereafter, whereas the increase in pressure rating continued until the 1960s" (Gordon 2004, p. 177).

sition to nuclear energy as a primary source of electrical power has occurred much more slowly than was anticipated. Political considerations contributed to the premature selection of the light-water power technology for commercial development and led to a path-dependent lock-in from which the U.S. commercial nuclear power industry has not yet been able to escape. It is possible that exploitation of renewable energy resources or development of one of the other alternative energy sources (possibly hydrogen) could over the next several decades emerge as a possible new general-purpose technology. However, none of the technical alternatives, including nuclear power, appear at present to promise sufficient cost reduction to enable the electric power industry to again become a leading rather than a sustaining source of economic growth in the U.S. economy (chapter 4).

In the late 1960s there were indications that mainframe computer development was approaching technological maturity. New trajectories were opened up by the development of the microprocessor. The minicomputer and later the microcomputer replaced the mainframe as the most rapidly growing segments of the computer industry and as important sources of output and productivity growth in the U.S. economy (Jorgenson 2001; Gordon 2004, pp. 22–49). Support by defense and space agencies contributed to advances in mainframe supercomputer speed and power into the early 1990s. But by the late 1990s substantial concern was being expressed about the sources of future advances in computer performance (National Research Council 1999; National Research Council 2003; chapter 5).

It would be premature to characterize either the Internet (chapter 6) or the space communication and earth-observing industries (chapter 7) as technologically mature. Both are, along with the computer and semiconductor industries, part of a rapidly evolving communications revolution that is expected to extend well into the first quarter of the twenty-first century.

A continuing issue in the field of information technology is whether productivity growth in the semiconductor industry can be expected to follow the classical pattern described by Evenson and Kislev. Moore's law has been interpreted to predict that the number of components per silicon chip in a microprocessor can be expected to double every eighteen months (chapter 5). But how long can this exponential rate of growth continue? Kenneth Flamm has argued that there is nothing inevitable about Moore's law. The law, at least into the 1990s, can be interpreted as "a self fulfilling expectations mechanism." Companies believed that their competitors were making R&D investments on the assumption that Moore's law was more or less valid and that they would have to make comparable investments to remain competitive: "Moore's law became an organizing and coordinating framework for private and public R&D in what is the largest and most globalized manufacturing industry in the world (Flamm 2004, p. 170; Branscomb 2005). I am reluctant, however, to embrace the view that self-fulfilling expectations, even in the field of information technology, can in the long run overcome the constraints imposed by basic physical principles. There have also been suggestions that Moore's law will face economic constraints before it confronts physical constraints—productivity gains from advances in processing power may not be worth the cost (Christensen, Anthony, and Roth 2004, pp. 162–165). The collapse of the communication industry "bubble" beginning in the late 1990s suggests some caution about the more extravagant expectations of continued logistic growth in productivity in the computer industry or the communications industry more generally.

In concluding this section let me again indicate why I have given so much attention to the issue of technological maturity in the general-purpose technologies discussed in this book. Historically, new general-purpose technologies have been the drivers of productivity growth across broad sectors of the U.S. economy. During the first half of the twentieth century, for example, productivity growth in the electric power industry was the major driver of productivity growth in the entire U.S. economy (chapter 4; Gordon 2004, pp. 22–49, 172–217). During the last several decades of the twentieth century the computer and microprocessor emerged as the major drivers of productivity growth in the U.S. economy (chapter 5; Jorgenson 2001).

It cannot be emphasized too strongly that if either scientific and technical constraints or cultural and institutional constraints should delay the emergence of new general-purpose technologies over the next several decades, the effect would surely be a slowing of productivity growth in the U.S. economy. Endless novelty in the technical elaboration of existing general-purpose technologies can hardly be enough to sustain a high rate of economic growth. In the case of the general-purpose technologies that emerged as important sources of growth in the U.S. economy during the last half of the twentieth century, it was primarily military and defense-related demand that drove these emerging technologies rapidly down their learning curves.

Structural Change

The second major issue I address in this chapter is whether changes in the structure of the U.S. economy and of the defense industries and the defense industrial base preclude military and defense-related R&D and procurement from continuing to play an important role in the generation of new general-purpose technologies.[6]

6. The defense industrial base includes the full range of industries that contribute importantly to the supply of products that make a unique or important contribution to the nation's defense capacity. The defense industry is a subset of the defense industrial base. It consists of firms that that produce largely or entirely for the military market. Thus, a firm in the steel industry that produces a specialized steel crucial for the production of submarines is part of the defense indus-

This issue has been the source of a substantial defense policy debate. Should, for example, military procurement policy be more explicitly directed to enhancing commercial technology spin-offs or to the development of dual technologies?[7]

Spin-Off

During the first two post–World War II decades, the spin-off issue attracted relatively little attention. It was generally taken as self-evident that substantial spin-offs of commercial technology could be expected from military procurement and defense-related R&D. It was also generally assumed that there was no need for policies to enhance the development of dual-use technology—technology developed with the specific objective of meeting both military and commercial demands: "Development of technology for commercial purposes typically consisted of self-initiated actions by private firms aiming at productivity and profits in the market economy. Development of technology for national defense (and defense-related missions) was often conducted in military laboratories and relied on federal funds to generate the technology needed by the Pentagon and other so-called mission agencies" (Alic et al. 1992, pp. 8–9).[8]

The slowdown in the rate of economic growth in the United States after the early 1970s generated considerable controversy about the role of the military in technology development.[9] Some critics had argued, even in the 1960s, that defense-related R&D was becoming a burden on economic growth. Military and space research was viewed as drawing scientific and technological capacity away from

trial base but not of the defense industry. In contrast, the missile and space vehicle industry, which sells 85 percent of its production to the military, is part of the defense industry, even though a small part of its production may be for civil uses (Flamm 2000, pp. 60–64).

7. Drawing on Cowan and Foray (1995), I define *spin-off* (or *spillover*) as a situation in which research or technology development conducted entirely for application in one sector (military or commercial) is readily applied or adapted for use in the other sector. In contrast, the term *dual purpose* is used to describe technologies that are specifically designed for use by, for example, both the military and the commercial sectors.

8. This "linear" view (figure 3.1) was articulated most forcefully in the influential report on postwar science policy, prepared at the request of President Franklin D. Roosevelt, under the direction of Vannevar Bush, *Science: The Endless Frontier* (1945). As late as the mid-1980s, the view was vigorously defended by George Keyworth, science advisor to President Ronald Reagan. Keyworth noted that the Reagan administration had been successful in increasing the share of federal research support for basic research while reducing the share devoted to applied research, technology development, and demonstration (Keyworth 1984).

9. The rate of growth of labor productivity (output per hour worked) in the U.S. nonfarm business economy declined from an annual rate of 3.5 percent in 1948–1973 to 1.5 percent per year in 1974–1995. It rose to 2.6 percent per year between 1996 and 2000, and to 4.0 percent between 2001 and 2004 (U.S. Bureau of the Census 1975; Council of Economic Advisors 2004, p. 342). For an analysis of the long-term productivity growth rate in the United States, see Jorgenson (2001) and Gordon (2004, pp. 22–49).

civilian applications. It was argued that the effect was to slow the advance of industrial technologies and reduce the rate of economic growth (chapter 1; Solo 1962, pp. 49–60; Kaldor 1981; Lichtenberg 1984; Dumas 1986; Lichtenberg 1989). In addition, defense industry firms, even those with substantial commercial activity, often failed to take advantage of technology transfer opportunities from their military R&D. Murray Weidenbaum (1992) has observed that "those defense firms that do operate in civilian markets often tend to maintain operationally separated insulated divisions that have little contact with each other, merely reporting to the same top manager" (p. 51).[10]

An effect of the productivity slowdown that began in the early 1970s was the questioning of the continuing relevance of the spin-off paradigm. The spin-off paradigm had emerged in an era "when the United States dominated world technology and national defense dominated United States technology development" (Alic et al. 1992, p. 25; box 8.1).

One of the factors advanced to explain the apparent loss of relevance of the spin-off paradigm was the emergence of economies in Western Europe and East Asia, particularly Germany and Japan, as competitors of the United States in a number of high-technology industries. Another was the revolution in military affairs associated with the winding down of the cold war. Reduction of tensions between the United States and the Union of Soviet Socialist Republics held out the promise, or at least the hope, for a dampening of the growth of military and defense-related budgets. A third was an intellectual challenge to the linear model of the role of science in technology development (see figure 3.1).

The linear model was deeply embedded in military procurement practice. Procurement followed a "pipeline" progression (see box 1.1). The first step is the preliminary process of concept definition, which itself requires substantial R&D resources, conducted by the contracting agency. The second phase involves design, demonstration, and validation. In this phase multiple firms may advance competing designs. The process then moves forward to full-scale prototypes and final demonstration—a phase that absorbs up to 20 percent of total system acquisition cost. This phase may involve more than a single firm. The final phase involves selection of a sole source contractor and development of the capacity for full production of the system. Modifications in concept, component technology, and even design

10. I first became sensitive to this issue in late 1970s when a student, then working for a Minneapolis-based firm that was substantially engaged in both defense and commercial R&D, noted in a term paper that a wall of secrecy separated the firm's military and commercial R&D. His interpretation was that the firm did not want its commercial division to be "contaminated" by the costly production processes involved in meeting defense-related contracting, quality, regulatory, and security requirements.

Box 8.1. Military R&D: The Productivity Puzzle

Is publicly funded military research and development (R&D) a source of technology development and productivity growth?[a] In earlier chapters I have presented historical case studies of the role of military procurement in technology development for six general-purpose technologies. The results seem to be in direct contradiction to a number of very careful econometric studies that show that measured private and social rates of return to military R&D have been very low and have had no discernible effect on industrial productivity growth in the United States (Lichtenberg 1984; Lichtenberg 1988; Lichtenberg 1989, p. 275).

There is a long history of studies of private and social rates of return to R&D. These studies trace back to the now classic studies of rates of return to agricultural research by Griliches (1958) and to industrial research by Mansfield and Beardsley (1977). The results of the large body of firm-level, sector-level, and economy-wide studies, combined with studies of the sources of productivity growth, have supported a view that the social rates of return to R&D have generally exceeded the rates of return on almost any other form of investment available to the U.S. economy. These high social rates of return contributed to a consensus that the United States was substantially underinvesting in R&D—and that this underinvestment was a substantial constraint on economic growth. Because of the spillover of R&D benefits in the form of consumers and producers surplus, even privately funded R&D shared the characteristics of public goods—the economic unit that generates the new technology can capture only a portion of the social benefits deriving from the research. The policy implication that has generally been drawn is that the United States should expand public-sector support for R&D to correct private-sector underinvestment.

The generality of this conclusion has been challenged, however, by studies by Lichtenberg (1984; 1988; 1989) and others that have attempted to measure the private rates of return to firms that conduct publicly funded research and technology development and of the firm and economy-wide spillover effects of such R&D. A large number of studies have failed to find significant private or social rates of return from publicly funded research conducted by private firms. However, "privately funded R&D in manufacturing industries is found to yield a substantial premium over the rate of return from 'own productivity improvements' derived from R&D performed with government funding" (David, Hall, and Toole 2000, p. 498).

It has been suggested that one explanation for these results may be that a high percentage of firm-level federally funded industrial research has been conducted by defense or defense-related firms. Neither the R&D nor the products resulting from such R&D are subject to a market test. The design of technol

(continued)

ogy, firm-level costs, and returns are heavily influenced by bilateral bargaining. Research results and technology development information are usually classified. A substantial share of the products derived from the federally funded R&D are often sold back to the government. Under these institutional arrangements conventional measures of profitability and productivity may not be appropriate (Griliches 1995, p. 82).

Tests conducted against the impact of federal funding on firm-level R&D or profitability do not, however, represent an adequate test of the effects of public sector, and particularly of defense-sector, R&D on economic performance. Public funding is often a complement (rather than a substitute) to private R&D and thus may enhance sector-level profitability and productivity. In addition, defense or defense-related procurement of services and products resulting from R&D may represent a substantial stimulus to firm-level research and technology development.

A number of early studies contributed to a presumption that much of the R&D conducted or funded by the public sector simply substituted for private-sector R&D—that it crowded out private-sector R&D. In an attempt to test the substitution hypothesis, David, Hall, and Toole (2000) conducted a critical review of the large body of econometric research studies that attempted to shed some light on the issue of whether public-sector R&D has been a substitute for, or a complement to, private-sector R&D. After sorting out the subset of studies that were adequately designed to test the substitution hypothesis, they found that the results from about one third were consistent with the substitution hypothesis, while two thirds were consistent with the complementarity hypothesis. I find it particularly significant that almost all of the higher-level aggregate studies were consistent with the complementarity hypothesis. The implication is that the more aggregate studies were able to capture elements of complementarity not detected at the individual firm level.

My own view is that we do not yet have, and perhaps cannot have, a body of rigorous econometric evidence against which to evaluate the economic impact of defense and defense-related R&D and procurement. David, Hall, and Toole explicitly eschewed any effort to assess the magnitude of the economic effect of complementarity. What are the implications for the attempt that I have made in this book to assess the significance of military procurement on the development of commercial technology? My answer is that careful narrative analysis of individual cases is at present a more effective method of capturing the effects of complementarity than econometric analysis. Paul A. David has also pointed out to me that narrative analysis may be better able to capture the long-term or lagged effects of public R&D investments (David 2004).

It is particularly important to assess the extent to which military procurement has induced both demand- and supply-side forces that have shortened

(*continued*)

Box 8.1. (*continued*)

the process of transition from initial concept to commercialization of new general-purpose technologies. The Semi-Automatic Ground Environment (SAGE) program (chapter 5) and the Apollo space mission (chapter 7) drove computer and microprocessor technology rapidly down their learning curves and advanced the development of commercial computer technology by at least a decade. Similar examples have been presented in chapters 2 through 7 of this book.

[a] I am indebted to Paul A. David and Frank R. Lichtenberg for comments on an earlier draft of this box.

often take place during this final stage. Numerous subcontractors will be involved in this stage. The demonstration-validation stage typically takes three to four years; the full-scale development phase, four to five years. The technology remains in production, on average, for about fifteen years and in service for many more (Rogerson 1994a; Flamm 1999, p. 241).[11]

By the mid-1980s this process was beginning to appear increasingly incongruent with R&D practice in the most advanced sectors of the commercial economy. The postwar U.S. economy had witnessed an accelerating transformation of the relationships among science, technology, and production. It became widely recognized that commercial production processes pressed more immediately, and sometimes beyond, advances in scientific knowledge (Alic et al. 1992, pp. 11–22).

Dual Use

Beginning in the mid-1980s and into the mid-1990s, civil-military industrial base integration, usually referred to in terms of development of "dual-use" military and

11. For an excellent brief history and critique of the weapons procurement process, see Lorell et al. (2000, pp. 15–18). Lorell and colleagues note that most past efforts to reform the weapons procurement process have had the effect of introducing regulations and processes that have raised barriers to the transfer of technology between military and commercial applications (Lorell et al. 2000, p. 15). For an exhaustive account of the evolution of military procurement policy and practice from the mid-1920s through World War II, see Holley (1964). For detailed analysis of the weapons acquisition process that retains substantial currency, see Scherer (1964) and Danhof (1968). For an economic approach to analysis of military procurement policy, see Rogerson (1994a; 1994b) and Sandler and Hartley (1995). For assessment of the economic and political effects of the privatization of procurement, see Markusen (2003).

commercial technology, became the conventional solution offered for how to solve the problems of rising cost and declining quality in post–cold war military procurement (Alic et al. 1992; Carnegie Commission 1993). Dual use was itself, of course, not new. Technology development in the several industries discussed in the earlier chapters of this book—in gun manufacture, aircraft, nuclear power, computers and semiconductors, and the space industries—has been, at least in part, dual use.

Efforts to institutionalize dual use in the procurement process have been controversial. Advocates believed that there were extensive dual-use opportunities in defense and commercial technologies, processes, and practices, and that U.S. Department of Defense (DOD) adoption of commercial acquisition practices could result in substantially lower procurement costs. Critics were skeptical about the extent of potential overlap between commercial and military products and processes. They argued that a specialized cadre of defense-related firms operating under close regulation and supervision remained the best solution for weapons procurement (Lorell et al. 2000, pp. 2–3).[12]

In the mid-1980s concerns about the competitiveness of a number of high-technology industries—semiconductors, for example—led to a proposal that the Advanced Research Project Agency (ARPA) be given authority to undertake commercial technology initiatives under the rubric of "dual use." In 1987 the Congress authorized a public-private Semiconductor Equipment Manufacturing Consortium (SEMATECH), in which the federal government proposed to contribute up to 50 percent of the cost over a five- to ten-year period for development of new high-technology equipment for the manufacture of semiconductors (chapter 6).

In an influential book published in 1992, *Beyond Spinoff*, five highly respected students of defense industrial policy urged a much more conscious weighting of potential civilian applications in defense-related research, technology development, and procurement (Alic et al. 1992). An implication is that support of defense-related technology should extend well beyond defense application in the way, for example, that ARPA continued to support the development of ARPANET even after the establishment of MILNET in 1982 (chapter 6).

The Clinton administration initially embraced, at least at the rhetorical level, the dual-use concept. It implemented a Technology Reinvestment Program (TRP), which had been approved by Congress but not implemented by the Bush admin-

12. In an iconoclastic book, *The Bottomless Well*, Huber and Mills (2005) argue that, though the relationship between military and commercial technology development can no longer be described by the dual-use and spin-off paradigms, the relationship remains important: "The military industrial complex now consists of two relatively thin bookends on our enormous civilian high-tech economy" (Huber and Mills 2005, pp. 149–150).

istration. The immediate objective of TRP was to advance the entry of new firms into the commercial development of technologies important to the military. Defense procurement reforms would be aimed at modifying military specifications to encourage greater use of products available in commercial markets. Military specifications were to be replaced by performance-based specifications (Perry 1994; Schmidt 2000). It was anticipated that lower costs arising out of economies of scale and other sources of cost reduction associated with commercial production would reduce acquisition costs.[13]

A second objective was to enhance the competitive position of U.S. manufacturers in high-technology products in international markets (Stowsky 1999).[14] The program received good marks for being well designed. It initiated a number of projects that were successful in meeting its dual-use objectives. A key difference between TRP and other dual-use programs was the emphasis that ARPA placed on the ability of TRP to generate technologies that resulted in commercial products. Other DOD dual-use efforts focused primarily on military applications and left subsequent commercial development of the technology largely to the private sector.

In retrospect it is clear that these efforts were badly underfunded and encountered substantial resistance from both the DOD and the large defense contractors.[15]

13. The dual-use initiative was proceeding simultaneously with a debate about defense industry diversification policy. Efforts were made by the DOD to encourage defense firms to respond to decline in DOD procurement resources by diversifying into production for the commercial market—mass transit vehicles in contrast to airframes, for example. The DOD motivation was to preserve the economic viability of enough defense industry firms to assure rivalry among firms in procurement. Some analysts were highly skeptical of the capacity of defense industry firms to successfully pursue diversification. Murray Weidenbaum (1992) argued, for example, that the managerial and technical capacities of most military contractors were best suited to making weapons and that their most appropriate response would be to downsize. Other analysts presented a much more optimistic perspective on defense conversion efforts and have criticized the Pentagon for not pressing conversion efforts more forcefully (Markusen and Yudken 1992; Gansler 1995).

14. In the 1980s and into the early 1990s, the Carter, Bush, and Clinton administrations were involved in the development of public institutions to advance commercial technology. Among the more effective were the Small Business Innovation Research Program (SBIR), the Advanced Technology Program (ATP) of the National Bureau of Standards and Technology, and the cooperative research and development agreement (CRADA) program (box 4.1). The ATP and CRADA programs involved cooperative agreements and cost sharing between private firms, government laboratories, and universities. I have discussed these and related programs in greater detail in Ruttan (2001, pp. 576–583). For the SBIR program, see Wessner (2000). For the ATP, see Wessner (1999). For a recent and very positive evaluation of CRADAs, see Adams, Chang, and Jensen (2003).

15. For a retrospective study of experiments by the Air Force to introduce commercial approaches in weapons procurement, see Lorell et al. (2000). Lorell and colleagues classify the opportunities for introduction of commercial approaches to procurement of military weapons or

The 1994 Republican Congress, as part of a general attack on federal technology programs, eliminated the TRP budget as of the end of the 1996 fiscal year (Morrison 1995). Once again, efforts to develop a "supply-side" or "technology-push" approach to strengthening incentives and capacity of firms in the defense industry in order to advance commercial technology had not met the test of political viability (Higgs 1994; Markusen 1997; Oden 1999; Stowsky 1999).

Consolidation

The demise of dual use as a major DOD initiative was confirmed in 1993 when Deputy Secretary of Defense William Perry announced an end to a half century of costly effort by the DOD to maintain rivalry among defense contractors by opposing mergers of firms producing comparable products (tanks, aircraft, satellites, submarines, and others). The Pentagon change in policy set off a flurry of mergers that reduced the ranks of the largest contractors (those with sales of over $1 billion each) from fifteen in 1993 to four in 1997 (figure 8.3).[16] The Pentagon permitted the contractors to write off the merger costs and have a return on investment, on the presumption that the mergers would save the government money in the future (Markusen 2000, p. 9).

At the beginning of the twenty-first century, the United States was still the dominant global producer of a broad range of capital- and skill-intensive defense and defense-related systems. It still accounted for more than two thirds of defense and defense-related R&D spending by the North Atlantic Treaty Organisation (NATO) countries and Japan. But the absolute size of defense procurement had declined in real terms to less than half of the 1985 cold war peak (Flamm 1999, p. 227; see figure 8.2). Furthermore, the share of output of the U.S. economy accounted for by the manufacturing sector had declined to less than 15 percent.

components into three categories: (1) pure commercial, (2) commercial but substantially modified for military use, and (3) military unique. The focus of the review is primarily on technologies and practices in category 2. Their conclusions are somewhat ambiguous. They find opportunities for substantial cost savings from adoption of commercial parts, technologies, and manufacturing and procurement practices in areas such as digital avionics. At the same time, they caution that careful oversight may be required to avoid sacrificing features essential to military performance in exchange for cost savings (Lorell et al. 2000, pp. 193–199).

16. "During the 1990s the number of credible aircraft prime integrators for fighters and bombers declined from seven to two. Similarly from 1990 to 1998 the number of U.S. missile manufacturers fell from fourteen to four while space launch vehicle producers declined from six to two. By 2001 only one credible developer of air to air missile producers remained active" (Lorell et al. 2002, p. 4). Similar consolidation occurred in Europe (Lorell et al. 2002, p. 6).

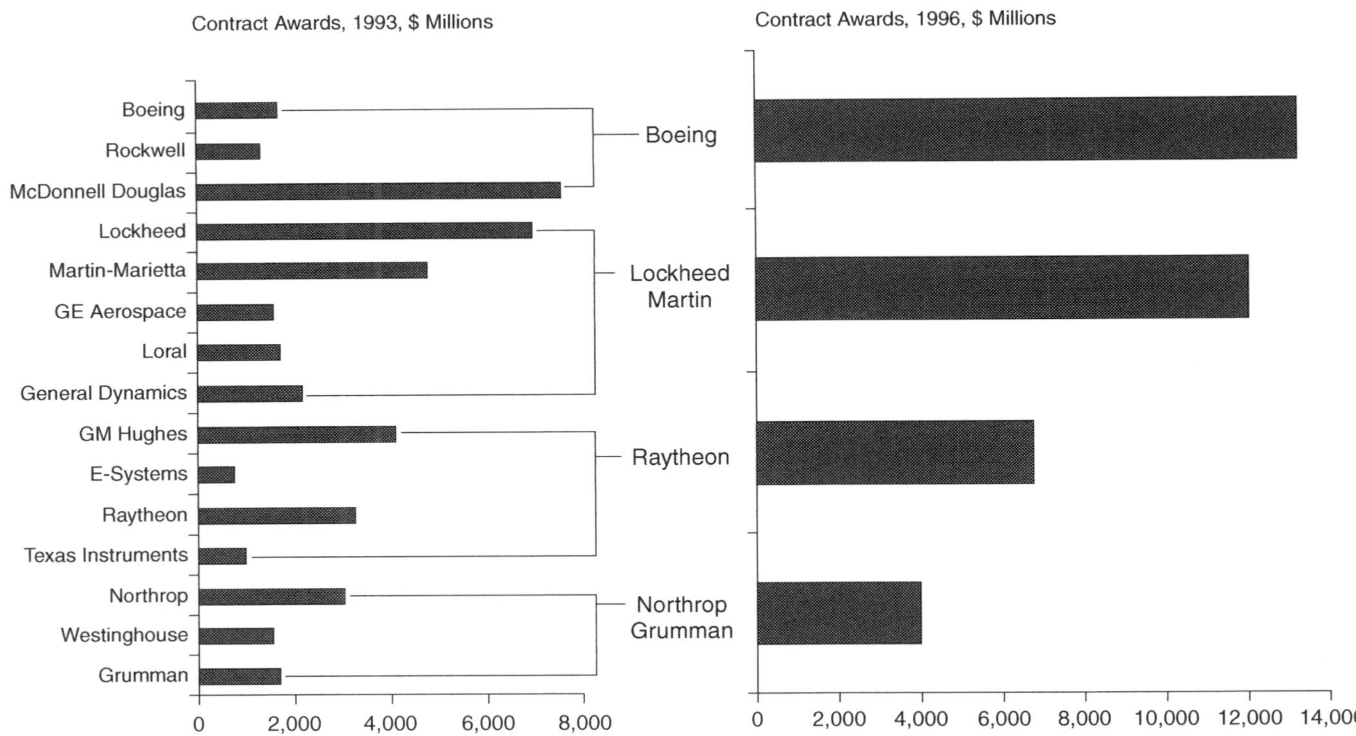

Figure 8.3. U.S. defense mergers in the 1990s. *Source:* Ann Markusen, 1998, "The Post–Cold War Persistence of Defense Specialized Firms," in *The Defense Industry in the Post–Cold War Era: Corporate Strategies and Public Policy Perspectives,* ed. G. I. Susman and S. O'Keefe (Amsterdam: Elsevier), 123. Copyright © 1998. Reprinted with permission of Elsevier.

Military and defense-related procurement had become a smaller share of an economic sector that itself accounted for a smaller share of national economic activity (Ruttan 2002).

An implication that a number of defense intellectuals drew from the structural changes in the defense industrial base and in the structure of the U.S. economy is that it no longer made sense to think of defense industrial policy in terms of a defense industrial base—a defense industrial base separate from the general industrial base. A policy issue that emerged from this discussion is the question, how should the United States proceed in construction of a transnational defense industrial-base strategy (Markusen 1998; Markusen 2000; Lorell et al. 2002)? In spite of extensive debate about the changing structure of the U.S. industrial economy and the changing structure of the defense industrial base, I have yet to identify a recent comprehensive analysis of the changing structure of the defense industrial base or of the policy implications for defense procurement.[17]

A Future for General-Purpose Technologies?

It is now time to turn to the third and even more difficult question. Military and defense-related R&D and procurement have played an important role in the emergence of most of the general-purpose technologies developed in the United States in the twentieth century. For more than a half century, the United States has been almost continuously engaged in either hot or cold wars. No matter how unpleasant, the question must be addressed: whether a major war, or threat of a major war, is necessary to induce the U.S. political and economic institutions to commit the very large resources necessary to generate or sustain the development of major new general-purpose technologies. In attempting to respond to this question, one must answer three additional questions.

17. For an early comprehensive study informed by the structure, conduct, and performance tradition of industrial organization research, see Gansler (1980). Gansler (1980, p. 4) argues that the defense industries and the defense industrial base were becoming both economically inefficient in the production of defense material and strategically unresponsive in terms of the ability to respond with sufficient speed to meet an emergency. Gansler insists that the solution would involve extending defense planning to incorporate the defense industries and the broader defense industrial base (pp. 263–264). The Gansler book was written before the consolidation of the defense industries in the 1990s. For useful perspectives on the changing structure of the defense industry, see Markusen (1998), Flamm (1999; 2000), and Lorell et al. (2002).

Private Sector

The first question is, can the private sector be relied on as a source of major new general-purpose technologies? The quick response is that it *cannot!* When new technologies are radically different from existing technologies and the gains from advances in technology are so diffuse that they are difficult to capture by the firm conducting the research, private firms have only weak incentives to invest in scientific research or technology development (Nelson 1959; Markiewicz and Mowery 2003). Each of the general-purpose technologies that I have reviewed have required several decades of public support, primarily in the form of military R&D and defense or defense-related procurement, to reach the threshold of commercial viability.[18]

Decision makers in the private sector almost never have access to the patient capital implied by a twenty-year, or even a ten-year, horizon (National Research Council 1999, pp. 233–235). This does not mean that the private sector cannot under the right conditions be a source of new general-purpose technology. When Mervin Kelly, director of research at Bell Laboratories, decided in the mid-1930s that vacuum tubes would become an obstacle to the efficient operation of telephone switchboards, he hired William Shockley to initiate a program to explore the potential of solid-state physics in communication technology (chapter 5).

Lewis Branscomb and colleagues at the Harvard Kennedy School of Public Affairs note, however, that in the United States many of the older research-intensive firms have almost completely withdrawn from the conduct of basic research and are making only limited investments in early-stage technology development.[19] During the first several decades after World War II, transient circumstances such as limited international competition or monopoly power reinforced by government regulation, as in the case of American Telephone and Telegraph, enabled research-intensive firms to take a long-term perspective on returns from basic research and

18. Referring specifically to the development of the electronic digital computer, Flamm (1988) notes that "the initial demonstration of radically new devices and architectural concepts were pioneered in an environment in which government (typically the military) shared the risks and costs. Development and refinement of the advances largely occurred in a commercial setting, as industry applied these ideas to more business-oriented applications" (p. 13).

19. Early-stage technology development includes "the technical and business activities that transform a commercially promising invention into a business plan that can attract enough investment to enter a market successfully and through that investment become a successful innovation" (Branscomb and Auerswald 2002, p. 1). Branscomb and Auerswald distinguish early-stage technology development from incremental or evolutionary technical change in the area of a firm's core business interests.

early-stage technology development. Since the early 1990s changes in industrial structure have driven many corporations that had previously been quite research intensive, such as Radio Corporation of America, International Business Machines, and Xerox, almost entirely out of basic research and early-stage technology development (Branscomb and Auerswald 2002; Auerswald et al. 2003).[20]

As support for basic and even early-stage technology development at large corporate laboratories has atrophied, new institutional arrangements for advancing scientific knowledge and technology development have emerged. These often involve complex relationships among large corporate research laboratories, specialized independent research laboratories, and venture capital "angels" or firms. These organizations have increasingly established close links with university or government laboratories that are involved in the more basic or conceptual investigations associated with technology development (Branscomb and Auerswald 2002, pp. 41–57).

The cases reviewed in this book suggest that entrepreneurial firms have often been most innovative when they have had an opportunity to capture the economic rents generated by complementary public investment in R&D. Even the most innovative firms often have great difficulty in pursuing more than a small share of the technical opportunities opened up by their own research, particularly if these opportunities fall outside the firms' core operating business lines. I find it difficult to anticipate that the private sector, without substantial public support for research and technology development, will become an important source of new general-purpose technologies over the next several decades.

Public Sector

A second issue is whether a more aggressive policy of public support for commercially oriented R&D might become an important source of new general-purpose technologies. I have argued in *Technology, Growth, and Development* (Ruttan 2001, pp. 368–422) that molecular biology and biotechnology will represent the source of the most important new general-purpose technologies of the early decades of the twenty-first century.

For more than three decades the molecular genetics and biotechnology research leading to the development of commercial biotechnology products in the phar-

20. Sustained public support has enabled the United States to remain preeminent in scientific research (Paarlberg 2004). Preeminence in scientific research is, however, only loosely linked to preeminence in technology development. See box 3.2 for a discussion of the importance of articulation between advances in scientific and technical knowledge. See also figure 3.1.

maceutical and agricultural industries was funded almost entirely by private foundations, the National Science Foundation, the National Institutes of Health, and the national energy laboratories—largely at government and university laboratories.[21] When the pharmaceutical and agricultural industries decided to enter the field in the late 1970s, they found it necessary to make very substantial grants and contracts to university laboratories to obtain a "window" on the advances in the biological sciences and in the techniques of biotechnology that were already under way in university laboratories (Ruttan 2001, pp. 377–384). When defense agencies in the United States and the Soviet Union (Union of Soviet Socialist Republics) began to explore the development of bioweapons and their antidotes, they also found it necessary to access capacities in molecular biology that were available only in university and health agency laboratories (box 8.2 Regal 2002).

I suggested earlier in this chapter that by midcentury a combination of concerns about environmental and energy security could induce sufficient public support for the development of alternative energy sources—sources other than carbon-based fossil fuels or nuclear energy. Modest efforts have been made since the mid-1970s to explore renewable-energy technologies, and considerable progress has been made in moving down the learning curves for photovoltaics and wind turbines (Alic, Mowery, and Rubin 2003, p. 12). The Bush administration has placed major emphasis on the potential of hydrogen technology to provide a pollution-free substitute for carbon-based fuels by the second half of the century (National Research Council and National Academy of Engineering 2004; Pacala and Socolow 2004; Romm 2004).[22]

It is possible that advances in scientific and technical knowledge will make possible the development of economically viable general-purpose energy technologies by the middle of the century; however, it would require major sustained public support for alternative-energy R&D, including the redirection of research programs of the national laboratories (chapter 4), to create the productive new opportunities

21. In 1938 Max Delbruck, at the California Institute of Technology, working with the support of the Rockefeller Foundation, identified DNA as the physical carrier of genetic information. In 1953 James Watson and Francis Crick, working at Cambridge University, identified the double helix structure of DNA molecules. In 1973 Stanley Cohen (Stanford) and Herbert Boyer (University of California, San Francisco), supported by National Institutes of Health grants, demonstrated a method for stably inserting genes from a foreign organism into a host genome (Ruttan 2001, pp. 373–374).

22. A hydrogen economy would involve "the production of molecular hydrogen using coal, natural gas, nuclear energy or renewable energy (e.g., biomass, wind, solar); the transport and storage of hydrogen; and the use of hydrogen in fuel cells, which combine oxygen with the hydrogen to produce electricity (and some heat)" (National Research Council and National Academy of Engineering 2004, p. 8; Huber and Mills 2005, pp. 75–90).

Box 8.2. Biotechnology and Bioweapons

As the cold war began to wind down in the 1980s, defense intellectuals and military planners began to advance a *spin-on* policy as a complement to policies designed to enhance the *spin-off* of military and defense-related technologies to commercial use.[a]

Three reasons were advanced for the shift toward spin-on. One was the growing availability of high-performance dual-use technology in commercial markets. Another was the high and increasing costs of developing new technologies intended primarily for defense applications. A third was that product life cycles in commercial markets were becoming shorter than in military systems. As a result the technological sophistication of commercial products was increasingly running ahead of military specifications (Samuels 1994, p. 28; Gansler 1995, pp. 135–146; Stowsky 1999).

The examples advanced by advocates of a more aggressive effort by military and defense-related agencies to draw on technologies available in commercial markets were typically specific rather than general purpose. Examples included technically advanced components of the Patriot missile produced by Japanese subcontractors, which were initially developed for commercial markets, antilock braking systems developed by the automobile industry to work rapidly and effectively in rugged environments, and computer-assisted surgical procedures that had initially been developed and applied in civilian settings.

Biotechnology provides a particularly compelling example of the spin-on of a major general-purpose technology from academic and civil government laboratories to both commercial and military applications (Henderson 1998; Alibek and Handelman 1999; Osterholm and Schwartz 2000; Miller, Engelberg, and Broad 2002). Major interest in bioweapons initially focused on the use of naturally occurring agents such as smallpox, anthrax plague, botulism tularemia, and hemorrhagic fever. Concern about the implications of bioweapons led in 1972 to an international Biological and Toxins Weapons Convention prohibiting the development and production of biological weapons. The convention was signed by the United States, the Soviet Union, and 138 other nations.

In 1969 President Richard Nixon ordered the U.S. bioweapons program discontinued except for research and development for defensive purposes. It is now known that until at least 1992 the Soviet Union was engaged in a massive bioweapons program that involved not only enhancement of the effectiveness of natural agents but also "the creation of genetically engineered strains of combination viruses that would defy conventional treatments" (Osterholm and Schwartz 2000, p. 41). It is an open question whether U.S. efforts to develop

(*continued*)

Box 8.2. (*continued*)

defenses against bioweapons did not also involve weapon development. To make a vaccine against such things as anthrax or salmonella, one must have access to supplies of the organisms.

The spin-on of civilian technology for military application in biotechnology opens up an entirely different set of policy issues than the spin-off of military technologies for commercial application. In the case of atomic energy, the U.S. Atomic Energy Commission exercised almost complete control over application development and diffusion of the technology for several decades (chapter 4). However, when civil and commercial development precedes military application, the possibility of achieving effective control over diffusion is greatly weakened.

In the case of bioweapons, knowledge of the technology needed to weaponize naturally occurring agents or to design new and more effective agents is widely available not only to scientists at university and government laboratories, but also more broadly to commercial laboratories and individual scientists. Little progress has yet been made in bringing the production and use of biological weapons under the discipline of international law (Meselson 2001).

[a] The term *spin-on* refers to the transfer of technologies initially intended for civil or commercial application to military or defense-related applications (Samuels 1994, p. 18).

for private-sector investment in alternative-energy technology development and diffusion. Thus, I am skeptical that alternative-energy technologies will become a low-cost source of economic growth.

Is it reasonable to anticipate the sustained public support that would be necessary to induce the development of new general-purpose commercial technologies? Traditionally, substantial public support for commercial technology development has been limited primarily to the fields of agriculture and health.[23] However, since the early 1980s the federal government has initiated several additional efforts to support commercial technology development. These include (1)

23. A reviewer raised the question whether there were any similarities that account for the success of public support for technology development in agriculture and health, and public support for technology development in the defense and defense-related industries. One response is that one cannot ignore the importance of very large public-demand-side programs in both agriculture and health.

public-private cooperative agreements designed to enhance the spin-off of technology from national laboratories in the form of cooperative research and development agreements (CRADAs), (2) an Advanced Technology Program at the National Institute of Standards and Technology (NIST) to provide financial support for public-private cooperation on R&D projects judged to have substantial public-goods dimensions or long-time horizons to achieve commercial viability, and (3) a Small Business Innovation Research (SBIR) program designed to support agency needs and to advance commercialization of technology.

The projects supported by these programs have generated a number of evolutionary advances in commercial technology and substantial economic benefits. The SBIA, which is funded by a formula that requires federal agencies that have R&D programs of more than $100 million to allocate at least 2.5 percent of their R&D budget support to SBIA projects, has received consistent support. The CRADA and NIST programs, however, have had difficulty sustaining political viability.[24]

Over two decades ago Richard Nelson (1982) observed that public support for the development of industrial technology was rarely forthcoming in the absence of an aggressive procurement program. The United States has not yet designed a coherent set of institutional arrangements for public support of R&D for civil purposes. My own preferred model for such institutions is the public-private research institutes directed to the problems of specific industries, such as the former National Committee on Aeronautics and the SEMATECH semiconductor equipment consortium (National Research Council 1999; Wessner 2000, pp. 129–130). When long-term political viability of public support for commercial technology development has been achieved, it has depended on vigorous support and representation by the industry for which the technology is being developed in the governance of the R&D organization.

In spite of a number of promising initiatives, I remain skeptical that public support for nonmilitary or defense-related technology development can be depended on to become the source of major new general-purpose technologies in the foreseeable future. These programs have generated substantial economic benefits, but even the most successful programs must be evaluated in terms of their contributions to evolutionary rather than revolutionary changes in technology.

24. The ATP budget grew from $67.9 billion in 1993 to $199.5 million in 1994. Its budget doubled in 1995 and was scheduled to double again in 1996. The growth of the program was curtailed, however, when it came under strong attack after the 1994 midterm election (Ruttan 2002, p. 379).

Is War Necessary?

A third question that must be answered is whether military and defense-related R&D can again become a source of major new general-purpose technologies. A negative answer to this question is already implicit in my discussion of the changing structure of the manufacturing sector of the U.S. economy, of the defense industries, and of the broader defense industrial base within the manufacturing sector.

Since the end of the cold war, the objectives of defense agencies have shifted toward enhancing their capacity to respond to shorter-term tactical missions. This trend was induced by an emerging consensus that the threat of system-level war ended with the cold war. Many defense intellectuals had come to believe that major interstate wars among the great powers had virtually disappeared by the end of the twentieth century (Barnett 2004, pp. 59–106). The effect has been to reduce incentives to invest in defense and defense-related "big science" and "big technology."

I remain somewhat skeptical, however, of the apparent consensus that supports these conclusions. A major problem in assessing technology futures is knowing what is going on *right now*. It seems quite apparent, for example, that if I had been writing this book in the mid-1970s, I would not have noticed, or would have attached little importance to, the commercial potential of research that had been supported by the ARPA Information Processing Office since the early 1960s. I certainly would not have anticipated the emergence of the Internet and its dramatic commercial and cultural impacts (chapter 6).

It would not have been unreasonable in the mid-1980s to anticipate that the massive scientific, technical, and financial resources, several multiples of the resources devoted to the Manhattan Project (chapter 4), committed to the Strategic Defense Initiative (SDI), popularly known as "Star Wars," could become the source of new general-purpose technologies.[25] The purpose of the SDI was to deter a potential attack on the United States by Soviet ballistic missiles armed with nuclear warheads. Design of the system would depend, for example, on advances in sensor, laser, computer, software, and guidance and control technologies. It was anticipated that an operational system might become available by 2004. It is not yet possible, however, to identify credible projections of major commercial spin-offs from SDI.[26]

25. For an early example of the exaggerated expectations of commercial technology spin-offs from the SDI, see Browne (1986).

26. In a 1989 report the Congressional Office of Technology Assessment indicated that the

It is quite possible, as in the case of the ARPA information technology program, that two decades is too short a time on which to base such assessments. There have yet been few published attempts to anticipate any radical new commercial technologies that might be induced by efforts to transform U.S. military forces to meet nontraditional and postconquest challenges, such as those encountered in Iraq (Deitchman 2004). In spite of these qualifications, however, I find it very doubtful, in the absence of at least a *threat* of major war, that the U.S. political system could be induced to mobilize the very large scientific, technical, and fiscal resources comparable to those required to initiate and sustain the development of major military and defense-related general-purpose commercial technologies of the past.

It was access to large and flexible resources that enabled powerful bureaucratic entrepreneurs such as Leslie Groves and Hyman Rickover (chapter 4), Joseph Licklider (chapter 6), and Del Webb (chapters 3 and 7) to mobilize the resources necessary to move the general-purpose technologies from initial innovation toward military and commercial viability (Doig and Hargrove 1987). They flourished in a political and administrative environment that accommodated their entrepreneurial energies—an environment that no longer exists for military and defense-related agencies and firms.

The rationalization of the processes involved in the allocation of resources to R&D in defense and defense-related procurement, combined with changes in the structure of the defense-related industrial base, has placed serious constraints on the ability of military R&D and defense-related procurement to continue to play a dynamic role as a source of new general-purpose commercial technologies.

Perspective

I have argued in this chapter that the U.S. private civil economy is unlikely to generate the major new general-purpose technologies necessary to sustain rates of productivity and economic growth comparable to the rates achieved during the early post–World War II decades and again during the information technology bubble that began in the early 1990s. I have also argued that in the absence of a major

SDI Innovative Science and Technology Office devoted its resources almost entirely to exploratory development work, with only informal consultation with the several military services (U.S. Congress, Office of Technology Assessment 1989, p. 54). For an exceedingly thorough early assessment of the scientific, technical, and policy considerations involved in the development and implementation of the SDI, see U.S. Congress, Office of Technology Assessment (1985). See also Haley and Merritt (1986), Fought (1987), Molina (1989, pp. 80–89), and Donohue (1994).

war, or threat of a major war, new general-purpose technologies are unlikely to emerge from military R&D and procurement.

Would even the large demands that would be placed on the U.S. economy by a major war or threat of war induce the development of major new general-purpose technologies comparable to the general-purpose technologies induced by World War II and the cold war? This is a question that I and other students of science and technology history and policy cannot answer. There can be no doubt, however, that the advances in scientific and technical knowledge and commercial technology induced by demand for defense and defense-related technology in the past imposed very heavy opportunity costs on the U.S. economy. I am reluctant to believe that the design of civil institutions capable of mobilizing the scientific, technical, and financial resources necessary to sustain rapid technical change and to direct these resources into economically and socially productive activities is beyond U.S. capacity. I am not, however, optimistic that we will design such institutions in the present economic and political environment.

I am left with three questions. Will it take a major war or threat of war to induce the mobilization of the scientific, technical, and financial resources necessary to develop major new general-purpose technologies? My answer to this question, based on historical experience, is that it may. But if the United States were to mobilize the necessary scientific, technical, and fiscal resources, would the U.S. defense industries and the broader U.S. defense industrial base be capable of responding? My answer is that the U.S. industrial base is losing its capacity to respond without drawing on international technical and financial resources. The third question is, would such an effort lead to the development of economically viable new general-purpose technologies? I remain skeptical.

When the history of U.S. technology development in the next half century is eventually written, it will focus on incremental rather than on revolutionary changes in both military and commercial technology. It will also be written within the context of slower productivity growth than the relatively high rates that prevailed in the United States during the 1950s and 1960s, and during the information technology bubble that began in the early 1990s.

ACKNOWLEDGMENTS

I am indebted to John Alic, Lewis Branscomb, Carl Pray, Robert Socolow, and Douglas Tiffany; to participants of the University of Minnesota Hubert H. Humphrey Institute of Public Affairs Freeman Center Policy Seminar (September 14, 2004); and to the History of Science and Technology Colloquium (December 3, 1994) for comments on earlier versions of this chapter.

REFERENCES

Adams, J. D., E. F. Chang, and J. L. Jensen. 2003. "The Influence of Federal Laboratory R&D on Industrial Research." *Review of Economics and Statistics* 85: 1003–1021.

Alibek, K., and S. Handelman. 1999. *Biohazard: The Chilling Story of the Largest Covert Biological Weapons Program in the World—Told from the Inside by the Man Who Ran It.* New York: Random House.

Alic, J. A., M. Branscomb, H. Brooks, A. B. Carter, and G. I. Epstein. 1992. *Beyond Spinoff: Military and Commercial Technologies in a Changing World.* Boston: Harvard Business School Press.

Alic, J. A., D. C. Mowery, and E. S. Rubin. 2003. *U.S. Technology and Innovation Policies: Lessons for Climate Change.* Arlington, VA: Pew Center for Global Climate Change.

Auerswald, P. E., L. M. Branscomb, N. Demos, and B. K. Min. 2003. *Understanding Private-Sector Decision Making for Early-Stage Technology Development.* Washington, DC: National Institute of Standards and Technology, Advanced Technology Program. Mimeograph.

Barnett, T.P.M. 2004. *The Pentagon's New Map: War and Peace in the Twenty-First Century.* New York: G. P. Putnam's Sons.

Branscomb, L. M. 2005. Letter to Vernon Ruttan. January 3.

Branscomb, L. M., and P. E. Auerswald. 2002. *Between Invention and Innovation: An Analysis of Funding for Early-Stage Technology Development.* Washington, DC: National Institute of Standards and Technology (NIST GCR 02-841).

Browne, M. W. 1986. "The Star Wars Spinoff." *New York Times Magazine* (August): 18.

Bush, V. 1945. *Science: The Endless Frontier.* Washington, DC: U.S. Office of Scientific Research and Development. Repr. National Science Foundation 1960, 1980.

Carnegie Commission on Science, Technology and Government. 1993. *New Thinking and American Defense Technology.* New York: Carnegie Commission.

Christensen, C. M., S. C. Anthony, and E. Roth. 2004. *Seeing What's Next: Using the Theories of Innovation to Predict Industry Change.* Boston: Harvard Business School.

Council of Economic Advisors. 2004. *Economic Report of the President Together with the Annual Report of the Council of Economic Advisors.* Washington, DC: U.S. Government Printing Office.

Cowan, R., and D. Foray. 1995. "Quandaries in the Economics of Dual Technologies and Spillovers from Military to Civilian Research and Development." *Research Policy* 21: 851–868.

Danhof, C. H. 1968. *Government Contracting and Technological Change.* Washington, DC: Brookings Institution.

David, P. A. 1990. "The Dynamo and the Computer: An Historical Perspective on the Modern Productivity Paradox." *American Economic Review* 80: 355–361.

———. 2004. Letter to Vernon Ruttan. April 11.

David, P. A., B. H. Hall, and A. Toole. 2000. "Is Public R&D a Complement or a

Substitute for Private R&D? A Review of the Econometric Evidence." *Research Policy* 29: 497–529.

Deitchman, S. J. 2004. "Completing the Transformation of U.S. Military Forces." *Issues in Science and Technology* 20: 62–68.

Doig, J. W., and E. C. Hargrove, eds. 1987. *Leadership and Innovation: A Biographical Perspective on Entrepreneurs in Government.* Baltimore: Johns Hopkins University Press.

Donohue, G. L. 1994. *Star Wars: A Case Study of Marginal Cost Analysis and Weapon System Technology.* Santa Monica, CA: RAND.

Dumas, L. J. 1986. *The Overburdened Economy.* Berkeley: University of California Press.

Evenson, R. E., and Y. Kislev. 1975. *Agricultural Research and Productivity.* New Haven, CT: Yale University Press.

Flamm, K. 1988. *Creating the Computer.* Washington, DC: Brookings Institution Press.

———. 1999. "Redesigning the Defense Industrial Base." In *Arming the Future: A Defense Industry for the 21st Century,* ed. A. R. Markusen and S. S. Costigan. New York: Council on Foreign Relations Press, 224–246.

———. 2000. "U.S. Defense Industries in the Post–Cold War: Economic Pressures and Security Dilemmas." In *The Place of the Defense Industry in National Systems of Innovation,* ed. J. Reppy. Ithaca, NY: Cornell University Peace Studies Program, Occasional Paper 25, 49–72.

———. 2004. "Moore's Law and the Economics of Semiconductor Price Trends." In *Productivity and Cyclicality in Semiconductors: Trends, Implications, and Question,* ed. D. W. Jorgenson and C. W. Wessner. Washington, DC: National Academies Press, 151–170.

Fought, S. O. 1987. *SDI: A Policy Analysis.* Newport, RI: Naval War College Press.

Gansler, J. S. 1980. *The Defense Industry.* Cambridge, MA: MIT Press.

———. 1995. *Defense Conversion: Transforming the Arsenal of Democracy.* Cambridge, MA: MIT Press.

Gold, D. 1990. *The Impact of Defense Spending on Investment, Productivity, and Economic Growth.* Washington, DC: Defense Budget Project.

———. 2005. "Does Military Spending Stimulate or Retard Economic Performance? Revisiting an Old Debate." Graduate Program in International Affairs, New School University, New York. Mimeograph.

Gordon, R. J. 2004. *Productivity Growth, Inflation and Unemployment: The Collected Essays of Robert J. Gordon.* Cambridge: Cambridge University Press.

Griliches, Z. 1958. "Research Costs and Social Rates of Return: Hybrid Corn and Related Innovations." *Journal of Political Economy* 66: 419–431.

———. 1995. "R&D and Productivity: Econometric Results and Measurement Issues." In *Handbook of the Economics of Innovation and Technical Change,* ed. P. Stoneman. Oxford: Blackwell, 52–89.

Haley, P. E., and J. Merritt. 1986. *Strategic Defense Initiative: Folly or Future?* Boulder, CO: Westview Press.

Henderson, D. A. 1998. "Bioterrorism as a Public Health Threat." *Emerging Infectious Diseases* 4: 488–492.

Higgs, R. 1994. "The Cold War Economy: Opportunity Costs, Ideology, and the Politics of Crisis." *Explorations in Economic History* 31: 283–312.

Holley, I. B., Jr. 1964. *Buying Aircraft: Material Procurement for the Army Air Forces.* Washington, DC: Department of the Army, Office of the Chief of Military History.

Huber, P. W., and M. P. Mills. 2005. *The Bottomless Well: The Twilight of Fuel, the Virtue of Waste, and Why We Will Never Run Out of Energy.* New York: Basic Books.

Jorgenson, D. W. 2001. "Information Technology and the U.S. Economy." *American Economic Review* 91: 1–32.

Kaldor, M. 1981. *The Baroque Arsenal.* New York: Wang & Hill.

Keyworth, G. A., II. 1984. "Four Years of Reagan Science Policy: Notable Shifts in Priorities." *Science* 224 (April): 9–13.

Lichtenberg, F. R. 1984. "The Relationship between Federal Contract R&D and Company R&D." *American Economic Review* 74 (2): 73–78.

———. 1988. "The Private R&D Investment to Federal Design and Technical Competition." *American Economic Review* 78: 550–559.

———. 1989. "Impact of the Strategic Defense Initiative on U.S. Civilian R&D Investment and Industrial Competitiveness." *Social Studies of Science* 19: 265–282.

Lorell, M., M. Kennedy, J. Lowell, and H. Levins. 2000. *Cheaper, Faster and Better? Commercial Approaches to Weapons Acquisition.* Santa Monica, CA: RAND.

Lorell, M. A., J. Lowell, A. M. Moore, V. Greenfield, and K. Vlachos. 2002. *Going Global? U.S. Government Policy and the Defense Aerospace Industry.* Santa Monica, CA: RAND.

Mansfield, E., and G. Beardsley. 1977. "Social and Private Rates of Return from Industrial Innovations." In *The Production and Application of New Industrial Technology,* ed. E. Mansfield. New York: Norton, 144–166.

Markiewicz, K., and D. C. Mowery. 2003. "The Federal Role in Financing Major Innovations: Information Technology during the Post War Period." Berkeley: University of California Haas School of Business. Mimeograph.

Markusen, A. 1997. "How We Lost the Peace Dividend." *The American Prospect* (July–August): 86–96.

———. 1998. "The Post–Cold War Persistence of Defense Specialized Firms." In *The Defense Industry in the Post–Cold War Era: Corporate Strategies and Public Policy Perspectives,* ed. G. I. Susman and S. O'Keefe. Amsterdam: Elsevier.

———. 2000. "Should We Welcome a Transnational Defense Industry?" In *The Place of the Defense Industry in National Systems of Innovation,* ed. J. Reppy. Ithaca, NY: Cornell University Peace Studies Program, Occasional Paper 25, 25–47.

———. 2003. "The Case against Privatizing National Security." *Governance: An International Journal of Policy, Administration and Institutions* 16: 471–501.

Markusen, A. R., and J. Yudken. 1992. *Dismantling the Cold War Economy.* New York: Basic Books.

Meselson, M. 2001. "Bioterror: What Can Be Done?" *New York Review* (20 December): 38–41.

Miller, J., S. Engelberg, and W. Broad. 2002. *Germs: Biological Weapons and America's Secret War.* New York. Simon & Schuster.

Molina, A. H. 1989. *The Social Basis of the Microelectronic Revolution.* Edinburgh, U.K.: Edinburgh University Press.

Morrison, D. C. 1995. "Technology Tussle." *National Journal* (1 April): 799–803.

National Research Council. 1999. *Funding a Revolution: Government Support for Computing Research.* Washington, DC: National Research Council.

———. 2003. *The Future of Supercomputing: An Interim Report.* Washington, DC: National Research Council.

———. 2005. *Measuring Research and Development Expenditures in the U.S. Economy.* Washington, DC: National Academies Press.

National Research Council and National Academy of Engineering. 2004. *The Hydrogen Economy: Opportunities, Costs, Barriers, and R&D Needs.* Washington, DC: National Academies Press.

Nelson, R. R. 1959. "The Simple Economics of Basic Scientific Research." *Journal of Political Economy* 67: 297–306.

———. 1982. "Government Stimulus of Technological Progress: Lessons from American History." In *Government and Technical Progress: A Cross-Industry Analysis,* ed. R. R. Nelson. New York: Pergamon Press, 451–482.

Oden, M. 1999. "Cashing In, Cashing Out, and Converting: Restructuring of the Defense Industrial Base in the 1990s." In *Arming the Future: A Defense Industry for the 21st Century,* ed. A. R. Markusen and S. S. Costigan. New York: Council on Foreign Relations, 74–105.

Office of Management and Budget, Executive Office of the President of the United States. 2004. *Historical Tables, Budget of the United States Government, Fiscal Year 2005.* Washington, DC: U.S. Government Printing Office.

Osterholm, M. T., and J. Schwartz. 2000. *Living Terrors: What America Needs to Know to Survive the Coming Bioterrorist Catastrophe.* New York: Random House.

Paarlberg, R. L. 2004. "Knowledge as Power: Science, Military Dominance, and U.S. Security." *International Security* 29: 122–151.

Pacala, S., and R. Socolow. 2004. "Stabilization Wedges: Solving the Climate Problem for the Next Fifty Years with Current Technology." *Science* 305 (August): 968–972.

Perry, W. J. 1994. "Specifications and Standards: A New Way of Doing Business." Washington, DC: Office of the Secretary of Defense, 24 June. Mimeograph.

Regal, P. J. 2002. *Bioterror: Scientists, Industry, Government—and Ethics.* Minneapolis, MN: The Humanist Institute.

Rogerson, W. P. 1994a. "Economic Incentives and the Defense Procurement Process." *Journal of Economic Perspectives* 8: 65–90.

———. 1994b. "Profit Regulation of Defense Contractors and Prizes for Innovation." In *Essays in the Economics of Procurement,* ed. A. G. Bower and J. N. Dertouzos. Palo Alto, CA: RAND National Defense Research Institute, 47–65.

Romm, J. J. 2004. "The Hype about Hydrogen." *Issues in Science and Technology Policy* 20(3): 74–81.

Rostow, W. W. 1956. "The Take-Off into Self-Sustained Growth." *Economic Journal* 66: 25–48.

———. 1960. *The Stages of Economic Growth: A Non-Communist Manifesto.* Cambridge: Cambridge University Press.

Ruttan, V. W. 2001. *Technology, Growth, and Development: An Induced Innovation Perspective.* New York: Oxford University Press.

———. 2002. "Can Economic Growth Be Sustained? A Post Malthusian Perspective." *Population and Development Review* 28: 1–12.

Samuels, R. J. 1994. *Rich Nation, Strong Army: National Security and the Technological Transformation of Japan.* Ithaca, NY: Cornell University Press.

Sandler, T., and K. Hartley. 1995. *The Economics of Defense.* Cambridge: Cambridge University Press.

Scherer, F. M. 1964. *The Weapons Acquisition Process: Economic Incentives.* Boston: Harvard University Graduate School of Business Administration.

Schmidt, C. P. 2000. *Changing Bureaucratic Behavior: Acquisition Reform in the United States Army.* Santa Monica, CA: RAND.

Solo, R. A. 1962. "Gearing Military R&D to Economic Growth." *Harvard Business Review*: 49–60.

Stowsky, J. 1999. "The History and Politics of the Pentagon's Dual Use Strategy." In *Arming the Future: A Defense Industry for the 21st Century*, ed. A. R. Markusen and S. S. Costigan. New York: Council on Foreign Relations, 106–157.

U.S. Bureau of the Census. 1975. *Historical Statistics of the United States: Colonial Times to 1970.* Washington, DC: U.S. Government Printing Office.

U.S. Congress, Office of Technology Assessment. 1985. *Ballistic Missile Defense Technologies.* Washington, DC: Office of Technology Assessment.

———. 1989. *Holding the Edge: Maintaining the Defense Technology Base.* Washington, DC: U.S. Government Printing Office, OTA-ISC-420.

Vernon, R. 1966. "International Investment and International Trade in the Product Cycle." *Quarterly Journal of Economics* 70: 190–207.

———. 1979. "The Product Cycle Hypothesis in a New International Environment." *Oxford Bulletin of Economics and Statistics* 41: 253–267.

Weidenbaum, M. 1992. *Small Wars and Big Defense: Paying for the Military after the Cold War.* New York: Oxford University Press.

Wessner, C. W., ed. 1999. *The Advanced Technology Program: Challenges and Opportunities.* Washington, DC: National Academy Press.

———, ed. 2000. *The Small Business Innovation Research Program: An Assessment of the Department of Defense Fast Track Initiative.* Washington, DC: National Academy Press.

Appendix 1

From Concept to Commercial Development: Seven General-Purpose Technologies

In this appendix, I list critical dates in the development of the several military or defense-related general-purpose technologies discussed in this book. I trace the time from the date of scientific discovery or technical invention to the date of economically viable commercial development. The time ranges from approximately fifteen to thirty years.

In 1968 Edwin Mansfield, in *Technological Change*, summarized the evidence available in the mid-1960s on the time interval between invention and innovation (or commercial adoption) for almost fifty technologies (pp. 74–77). With few exceptions the period from invention to innovation was less than ten years. He also presents data suggesting that the time from invention to innovation declined from the early twentieth century to the post–World War II period.

1. Interchangeable Parts

1798 Bill authorizes President George Washington to establish public arsenals.

1815 Roswell Lee is appointed superintendent of Springfield Armory.

1818 Roswell Lee contracts with Thomas Blanchard to develop irregularly shaped gun stocks.

1819 John Hull is appointed assistant armorer at Harpers Ferry Armory.

1823 First guns with fully interchangeable parts are delivered to the U.S. Army by Harpers Ferry Armory.

1834 Guns with fully interchangeable parts are produced at two different armories for the first time.

2. Jet Propulsion

Mid-1930s Germany and Britain initiate jet aircraft development programs.

Early 1940s	U.S. National Committee on Aeronautics establishes a Special Committee on Jet Propulsion. Westinghouse and Allis Chalmers initiate jet engine development.
1941	U.S. Army decides to put Whittle-designed engine in a jet engine to be built by General Electric for an aircraft to be built by Bell Aircraft.
1948	First commercial jet airliner, powered by four Rolls Royce engines, built by Vickers.
1953	First de Havilland Comet begins scheduled service between London and Johannesburg.
1954	Lockheed introduces L-188 Electra passenger jet powered by Allison 581 engine.

3. Nuclear Power

1933	Otto Hahn and Fritz Strassman at Kaiser Wilhelm Institute in Berlin split atoms by bombarding nuclei with neutrons.
1942	Manhattan Engineering District is established to develop atom bomb.
1942 (Dec.)	Enrico Fermi demonstrates controlled nuclear fusion at University of Chicago Stag Field laboratory.
1945 (Aug.)	Uranium bomb is dropped over Hiroshima; plutonium bomb is dropped over Nagasaki.
1955	*Nautilus*, first nuclear submarine, is launched.
1956	First demonstration nuclear power plant at Shippingport, Pennsylvania, is placed on line.
1962	Yankee and Consolidated Edison commercial-scale nuclear power plant is put on line.

4. Electronic Digital Computer

1937	First electronic digital computer is designed by John Atanasoff at Iowa State University; it is demonstrated by Atanasoff in 1940.
1946	John W. Mauchly and J. Prosper Eckert Numerical Integrator and Calculator is demonstrated.
1952	International Business Machines (IBM) commercial version of Defense Calculator, IBM 701, is placed on the market.
1953	Engineering Research Associates deliver Atlas computer to National Security Agency.

1957 Whirlwind computer is developed at the Massachusetts Institute of Technology (MIT); it is produced by IBM for U.S. Air Force Semi-Automatic Ground Environment (SAGE) project.

1958 First section of SAGE project becomes operational.

5. Semiconductors

1936 William Shockley hired by Bell Laboratories to initiate solid-state research program.

1947 Shockley, John Bardeen, and Walter Brattain produce first point-contact transistor.

1958 Jack Kilby of Texas Instruments develops first integrated circuit. Robert Noyce and Gordon Moore of Fairchild Conductor invent planar process integrated circuit.

1970 First microprocessor invented at Intel.

6. Internet

1962 Defense Advanced Research Projects Agency (DARPA) Information Processing Office (IPO) is established. Joseph Licklider is appointed first director.

Late 1950s–early 1960s Paul Baran, of the RAND Corporation, writes a series of papers proposing packet routing and switching. Similar proposal is made by Donald Davies of the British National Physics Laboratory in the mid-1960s.

1966 Lawrence Roberts is given mandate to build large computer network by IPO Director Robert Taylor.

1969 Bolt, Beranek and Newman complete work on development of the first Interface Message Processor (IMP) designed to route message packets along alternative routes.

1972 Internet is demonstrated at International Conference on Computer Communication.

1990 Tim Breners-Lee, then working at the European Center for Nuclear Research (CERN) in Switzerland, creates the first server, browser, and protocols that have become central to the operation of the World Wide Web.

1994 Netscape, founded by Marc Andreesen, introduces the first easy to use commercial browser.

7. Satellites

1898–1916	Theoretical and experimental work by Konstantin Tsiolkovsky and Robert Goddard demonstrates feasibility of rocket flight in space.
1940–1944	Development and deployment of V-2 rockets in Germany by team is directed by Werner von Braun.
1945	German rocket team surrenders to U.S. Army. Team is brought to Fort Bliss, Texas, to continue experimental work.
1949	Rocket team is transferred to Redstone Arsenal in Huntsville, Alabama. The "super V-2" Redstone rocket is developed.
1955	President Eisenhower approves plans to develop satellite as part of U.S. participation in International Geophysical Year. Project Vanguard is assigned to U.S. Navy.
1957	The Soviet Union launches *Sputnik I* and *II*.
1958	SCORE, first active communication satellite, launched by U.S. Army. Transmitted Eisenhower Christmas message to the world.
1959	After several failures the U.S. Army successfully launches Vanguard III.
1960	Weather Bureau Television and Infra-Red Observation Satellite is launched.
1963	COMSAT Act authorizes commercialization of communication satellites.
1972	First Landsat earth-observing satellite is launched by the National Aeronautics and Space Administration (NASA).

Appendix 2

Computers, Microprocessors, and the Internet: A Counterfactual History

In this appendix I present an abbreviated counterfactual history of the electronic digital computer, the microprocessor, and the Internet. In constructing the counterfactual narrative, I assume a world in which military and defense-related research and technological development and procurement have played no role in the development of the three interrelated technologies. I have, in effect, assumed a world in which World War II and the cold war either did not occur, or at least had no impact on commercial technology development. The timing of events in the narrative is clearly subjective. But it does draw on my study of the development of these three technologies (see chapters 5 and 6). I am indebted to Jeffrey Yost for comment on an earlier version.

1940	First electronic digital computer is demonstrated by John Atanasoff at Iowa State University.
1947	Point-contact transistor is invented by Shockley, Bardeen, and Brattain at Bell Laboratories.
1960	First commercial electronic digital computer is introduced.
1965	First commercial application of transistor is effected.
1968	Planar process integrated circuit is invented.
1975	Integrated circuits begin to replace vacuum tubes in telephone switchboards and computers.
1980	Minicomputer is introduced.
1983	Microcomputer is introduced.
1985	National Science Foundation initiates support for development of software to enable computers of different designs to "speak to each other."
1992	Computer Interface Message Processor is invented.

2002	Internet browser is invented.
2004–2006	Rapid diffusion of personal microcomputers takes place.
2010	Measurable impact of computer on total factor productivity in private business sector is detected.

Author Index

Page numbers followed by n indicate notes.

Subject Index

Page numbers followed by n indicate notes.

CPSIA information can be obtained
at www.ICGtesting.com
Printed in the USA
BVHW04*0518180718
521886BV00007B/137/P

9 780195 188042